W9-BJF-790

Evaluating Radiographic Quality

The Variables and Their Effects

Evaluating Radiographic Quality

The Variables and Their Effects

O. Gary Lauer, Ph.D., RT (R) ARRT
John B. Mayes, M.A., RT (R) ARRT
Rosemary P. Thurston, M.S., RT (R) ARRT

THE BURNELL COMPANY/PUBLISHERS, INC.
Mankato, Minnesota 56002

Published by
The Burnell Co./Publishers, Inc.
821 North Second Street, P.O. Box 304
Mankato, MN 56002
ISBN 0-916973-04-2

First reprint 1992.

Dedication

The Army Radiographer plays a critical role in conserving the fighting strength of America's armed forces during war and peacetime. The doubts and indescribable perplexities facing Army radiographers during the Spanish-American War, World War I, World War II, Korean War, Vietnam War, Persian Gulf War and the periods of peace in between, did not become the lot of those Army radiographers who followed in their footsteps.

We dedicate this textbook, therefore, to the more than 20,000 medical radiographers that were educated and trained by the United States Army since 1898, the period of the Spanish-American War, when X-rays were first used by the United States Army. Their quest for the production of quality medical radiographic images, under often extremely adverse conditions throughout the world, is expressed in the motto of today's Army radiographer:

"STRIVE FOR IMAGING EXCELLENCE"

Acknowledgements

The authors of this textbook gratefully acknowledge the assistance of Rita J. Robinson, B.S., RT (R) ARRT, Director of the Radiography Program, Memorial Hospital System, Houston, Texas, for producing the majority of the radiographic illustrations presented in this textbook and Mark Beatty for producing the line-drawn illustrations. Also, the authors gratefully acknowledge the tremendous support and assistance given by Mr. R. B. Phillips and Mrs. Elda Dubke, President of the Burnell Company/Publishers, Inc., Mankato, Minnesota, in the production of this quality textbook.

About The Authors

O. Gary Lauer, Ph.D., RT (R) ARRT received his Associate f Arts in Radiologic Technology and entry level radiography aining at Santa Barbara City College, Santa Barbara, California, achelor of Science in Radiologic Technology and Master of ducation in Educational Administration from the University of levada at Las Vegas and Doctor of Philosophy in Education from e University of Santa Barbara. He is registered by the American egistry of Radiologic Technologists.

During the past two decades, he has served in staff and supervising radiography capacities among several hospitals and clinics nd held radiologic technology program director and instructor ppointments at Laramie County Community College, Cheyenne, Wyoming and Casper College, Casper, Wyoming. He acepted a direct commission as a Captain in the United States Army nd was appointed to three years of active duty at the Academy of lealth Sciences, Fort Sam Houston, Texas, as Chief of the X-Ray ranch and Program Director of the United States Army X-Ray pecialist Program.

Upon returning to civilian life, Gary served as product manager or teleradiology systems at Image Data Corporation, San Antoio, Texas. Today, Gary is a human development consultant edicated to improving the level of creative problem solving mong Americans.

His professional contributions to radiologic technology include ast president of the Western Intercollegiate Consortium on ducation in Radiologic Technology, WICERT, previous memerships in national and state radiologic technology societies, the uthor of numerous articles on radiography education and history, e writer of teleradiology operating manuals and coordinating ditor and contributing author for a multi-authored textbook about ollegiate radiography educational administrative practices.

Gary's contributions to *Evaluating Radiographic Quality* inlude that of coordinating editor and the writing of chapters 1, 2, , 4, 8, 17 and the Appendix.

John B. Mayes, M.A., RT (R) ARRT received his Bachelor f Arts in Physical Education and Health for Men from the University of Northern Iowa and Master of Arts in Management and Human Relations from Webster University. He is an Army trained radiographer, registered by the American Registry of Radiologic Technologists and licensed by the State of Texas.

During the past two decades, John has been a civilian radiologic technology instructor at the United States Army X-Ray Specialist Program, Academy of Health Sciences, Fort Sam Houston, San Antonio, Texas. He and His wife, Deborah, are the parents of two children, Sara and Adam. His hobbies are computer programming and woodworking.

John's contributions to *Evaluating Radiographic Quality* include the writing of chapters 5, 7, 9, 10, 11, 13 and the Index.

Rosemary P. Thurston, M.S., RT (R) ARRT received her Associate of Applied Science degree at St. Philip's College and Bachelor of Science and Master of Science degrees at Southwest Texas State University. She received her radiography training at Methodist Hospital, Indianapolis, Indiana, and radiation therapy technologist training at the M.D. Anderson Cancer Research Center, University of Texas, Houston, Texas.

She is registered by the American Registry of Radiologic Technologists and Licensed by the State of Texas. In addition to raising her family, she served as radiography instructor at Methodist Hospital, Indianapolis, Indiana, technologist and supervisor of X-Ray Therapy and Nuclear Medicine at the Community Hospital in Indianapolis, x-ray therapy technologist and dosimetrist at Brooke Army Medical Center, Fort Sam Houston, Texas and instructor for the United States Army X-Ray Specialist Program, Academy of Health Sciences, Fort Sam Houston, Texas. Currently, Rosemary is an instructor in the Faculty Development Branch at the Academy of Health Sciences. Her professional contributions to radiologic technology include past president of the Alamo Society of Radiologic Technologists and memberships in the Association of Educators in Radiological Sciences and Texas Society of Radiographers.

Rosemary's contributions to *Evaluating Radiographic Quality* include the writing of chapters 6, 12, 14, 15, 16 and the Index.

Table of Contents

Foreword xiii
Preface xv
1 The Early Days of Radiography 1
Plate I: Principal Properties of Radiographic Quality ..... preceding page 17
2 The Photo-Radiographic Properties of Image Visibility 17
3 The Geometric Properties of Image Formation 27
4 The Perceptual Properties of Image Quality 35
Plate II: The Effects of Principal Exposure and Imaging Factors on Radiographic Quality ..... preceding page 43
The Variables
5 The Effects of Kilovoltage on Radiographic Quality 43
6 The Effects of Milliampere-Seconds on Radiographic Quality 49
7 The Effects of Distance on Radiographic Quality 57
8 The Effects of Focal Spot Size on Radiographic Quality 63
9 The Effects of Filtration on Radiographic Quality 69
10 The Effects of Beam Restrictors on Radiographic Quality 75
11 The Effects of Tube-Object-Film Alignment on Radiographic Quality 81
12 The Effects of the Patient on Radiographic Quality 87
13 The Effects of Grids on Radiographic Quality 93
14 The Effects of Intensifying Screens on Radiographic Quality 101
15 The Effects of Motion on Radiographic Quality 109
16 The Effects of Film Processing on Radiographic Quality 115
17 Radiographic Exposure Conversion Problems 121
Appendix–Significant Events Leading Up to and Beyond the Discovery of X-rays 133
Index 137

Foreword

). Gary Lauer, Ph.D., John B. Mayes, M.A. and Rosemary P. urston, M.S., all of whom are registered radiographers with the nerican Registry of Radiologic Technologists, have produced elcomed edition which explains at varying levels of detail the nplex physical parameters affecting the image quality of diag- stic radiographs. Knowledgeable and experienced radiogra- ers and physicians should recognize the excellence of their rk because it is a comprehensive, focused and efficient evalu- on of the important exposure variables on image quality.

'or the professional, this text is a new and refreshing approach presenting clear and concise descriptions of often confusing l complex terminology, definitions and concepts concerning cause and effect relationships of radiographic imaging. The ormation is easy to find and easy to read. For the fledgling dent radiographer and physician (to include resident radiolo- ts), this text is a "valuable gem" due to the comprehensive cational format that enhances learning.

The authors' vast clinical and educational experience in diag- stic radiography has combined to integrate all subjects accord- to a solid educational foundation. So, even though this book ttractive as an efficient "quick" reference, it is also an outstand- educational tool specifically related to a formal course on liographic film evaluation in the professional diagnostic imag- curriculum for technologists and physicians alike.

This text goes beyond focusing solely on the traditional radiog- hy qualities of contrast, density, detail and distortion. Lauer, ayes and Thurston have done something innovative, unique and remely effective by grouping the various radiographic quali- s and then comparing and evaluating each exposure variable to h of these qualities. For example, grids, a major exposure riable is compared separately to contrast, density, detail and dis- tion.

The authors use numerous illustrations, (quite a few of which archival), and diagrams that complement and clarify the con- ots described in the text. This correlation between text and grams and illustrations is excellent. At the end of the text, for h major imaging parameter, the authors have provided a mprehensive summary of those physical factors which affect liographic quality by using a comparative system of +, -, and 0 symbolism for more, less and no change, respectively. This shows the degree to which the physical factors affect the major imaging parameter as it relates to each radiographic quality.

Each chapter begins with a forwarding chapter outline, a list of learning objectives, an easily comprehensive presentation and summary of content, a set of summary review questions and a detailed bibliography. Thoughout the content of the book, the authors include the development of the logic for solutions rather than just the answer in order to assist the reader in learning how to solve the math.

The inclusion of three additional chapters relating to the early days of radiography, analyzing the perceptual properties of radio- graphic quality and applied exposure conversion problems de- picting real work experiences provide invaluable insights that are not found in any other similar book. The historical chronology of events leading up to and beyond the discovery of the x-ray supplied in the appendix is fascinating and complements the historical perspective on film evaluation provided in the first chapter, "The Early Days of Radiography."

Gary, John and Rosemary have gone to great lengths to simplify the complexities associated with the evaluation of radiographic quality into a highly effective and efficient reference and learning tool; one that will become a standard by which all others will be judged. As an experienced health physicist, I highly recommend this book for every science professional and medical imaging educational program with the important responsibility of manipu- lating or teaching the manipulation of x-ray exposure factors in the production of high quality diagnostic radiographic images. The knowledge and skills to be gained from *Evaluating Radio- graphic Quality*, will improve the quality of patient care by increasing the life span of expensive x-ray equipment and sup- plies and reducing unnecessary exposure levels to both patient and operator.

John H. Bell, Ph.D.
Associate Professor of Radiological Sciences
University of Nevada, Las Vegas

Preface

The ability of the radiographer to analyze the diagnostic quali-
of a radiograph is at the heart of insuring optimum radiography
atient care since the production of good quality radiographs helps
e physician to render the best possible radiographic interpreta-
on of the medical condition of a patient. Mastering the fun-
amentals of radiographic film evaluation, therefore, is one of
e most important and complicated tasks facing students of
edical radiography. Because this task is analytical in nature,
demands that the student have prerequisite knowledge and com-
rehension in most of the basic radiography subjects, i.e.,
diographic physics, anatomy and physiology, patient care,
edical terminology, radiographic exposure, film processing,
diographic positioning, radiation protection, etc.

Although written primarily as an instructional tool for students
nrolled in the professional medical radiography curriculum,
valuating Radiographic Quality will be beneficial as a specific
eference for practicing radiographers, radiologists, physicians
volved with the use of diagnostic x-rays in their medical, den-
l, veterinary, podiatry, or chiropractic practices. Radiology
esidents and other students of the health science professions who
equire the capability of determining the diagnostic quality of
diographic films should have this volume. Also, this textbook
hould be of tremendous benefit to medical imaging graduates
ho are preparing for national certification or state licensure
xaminations.

Five dimensions have been carefully designed into the produc-
on of this textbook. The first dimension emphasizes the book
s a functional teaching and learning tool and not just another
andard reference adapted for the bookshelf after classroom in-
ruction has ended. Each chapter is educationally sound and is
uided by specific educational objectives that cover the major
bject areas. For chapters 5 through 16, a basic review is pro-
ided of the role of the principal exposure variables. End-of-
hapter summaries are followed by study questions based on
arning objectives. Bibliographies are provided so that students
nd educators have other reference sources available for addi-
onal inquiry concerning chapter contents. This textbook,
erefore, may be used effectively by individual students for self-
rected learning as well as by didactic and clinical instructors
ho plan and guide the formal learning experience in the classroom, laboratory, or clinical environment.

The authors recommend that this book be adopted specifically for a formal course in the radiography curriculum entitled, Radiographic Film Evaluation. There is sufficient coverage of material to structure a 3- or 4-semester lecture (lab optional) credit emphasis.

Much of the footwork has already been done for the instructor and student since educational objectives and study questions are provided on correlated chapter contents. This leaves room for creativity in terms of what the teacher and learner want to do with organization and implementation of the subject matter.

Although this book is at the level of analysis and draws heavily on previous knowledge and comprehension levels of radiography, there is sufficient depth of coverage on critical aspects concerning the role of key exposure variables in radiography to make the book attractive as an important adjunct to the various textbooks students are required to obtain for other radiography courses.

The second dimension deals with an historical account of the early days of radiography. Chapter 1, entitled, "The Early Days of Radiography," explains original methods early radiographers employed in producing images and assessing and manipulating radiographic film quality along with additional important background information. Many technical issues faced by early radiographers were surprisingly similar to those faced by radiographers of today. Much of what happened during these early days helped shape the profession of radiography as we know it today. Students should know this. Knowledge of the past gives particular meaning to the present and future. This also helps instructors to generate student interest considering that a subject like radiographic film evaluation has the potential for being rather dry to the nonscientifically or technically motivated student.

The third dimension focuses on developing a meaningful conceptual foundation of radiographic quality. Chapters 2, 3, and 4 introduce the photo-radiographic, geometric, and perceptual properties of radiographic quality and provide knowledge and comprehension about their principle influencing factors. Chapter 4, entitled, "The Perceptual Properties of Image Quality" presents original information concerning radiographic quality not covered by most textbooks. The student learns that radiographic quality may be influenced by extrinsic "perceptual" conditions in addition to intrinsic conditions of "image visibility and form." This chapter concludes with a detailed list of criteria designed to objectively assist in radiographic film evaluation, a subject which has not been given much formal emphasis in the literature and medical radiography curriculum, that is, until now.

With knowledge and comprehension about image visibility, image formation, and the quality of image details, the student is ready to learn how to apply the intricate details of the relationship between a single principal exposure and imaging variables, effect on the radiographic qualities. This fourth dimension, incorporated into Chapters 5 through 16, goes beyond simply defining radiographic properties and corresponding influencing variables as presented in Chapters 2, 3, and 4. The study of the effect of a single principal exposure variable on radiographic film quality provides considerable order and intensity in learning how to objectively evaluate the diagnostic parameters of radiographs.

The fifth dimension of this textbook concentrates entirely on the learning level of analysis. Students are provided with opportunities to apply their new analytical capabilities on word problems reflecting technical alterations to hypothetical imaging situations. Progressively more difficult multiple-type conversion problems and solution strategies are presented. We encourage all instructors and students to carry the emphasis of this chapter further by developing their own set of exposure conversion problems using the strategies employed in Chapter 17. This is one of the most comprehensive coverages yet to be published on strategies for manipulating exposure and imaging variables for the maintenance of diagnostic film quality.

The authors based the contents of this textbook on valid and reliable sources of information. No intention was made to report on every literary to technological investigation or implication currently underway in the radiography industry or profession since the fundamentals of static radiographic imaging are stable and have not significantly changed since 1923, a time in which the technology of medical radiography is considered to have met all the modern prerequisites. This would have been an unrealistic expectation and would have required a volume considerably different from this one.

We are confident that adaptation to future technological changes in conventional radiographic imaging should be easily achieved if students learn the fundamentals of radiographic film evaluation as presented in the enclosed chapters. For example, as the technology shifts toward digital imaging, the term "Focal Film Distance, FFD," will increasingly be replaced with "Source to Image Distance, SID." Thoughout this book, we have used both terms interchangeably. The scope of coverage, and new material presented in this book provides an important and much needed contribution to the radiography literature in the area of film evaluation. We sincerely hope that it helps to make mastering the fundamentals of radiographic film evaluation an objective, relatively simple and more enjoyable experience.

O. Gary Lauer,
John B. Mayes
Rosemary P. Thurston
San Antonio, Texas.

Chapter 1

The Early Days of Radiography

Chapter Outline

1. Learning Objectives
2. Introduction
3. Roentgen's Discovery
4. The Period of X-ray Commercialization: 1895-1923
5. The Birth of the Radiography Profession
6. Early X-Ray Machines
7. Early X-Ray Tubes
8. Coolidge X-Ray Tubes
9. Early Recording Mediums
10. Early Exposure Techniques
11. Early Processing Methods
12. Early Radiation Protection Guidelines
13. Summary
14. Study Questions
15. Bibliography

Learning Objectives

Upon completion of this chapter the student should be able to:

1. List the date, specific place and general time of day the discovery of x-rays took place by Dr. Roentgen.
2. Give the range, in years, encompassing the period known as that of x-ray commercialization.
3. Give the name of the roentgen pioneer credited as the first to establish a commercial x-ray laboratory in Chicago in May, 1896.
4. List three examples of terms used to describe then what today is known as radiography.
5. Give the name of the individual credited with organizing early American x-ray operators into a professional association known as the American Association of Radiological Technicians and explain why this association was formed.
6. List the organizational dates of the American Association of Radiological Technicians and the American Registry of Radiological Technicians.
7. Give the name of the first individual to be certified by the American Registry of Radiological Technicians.
8. List the names given to the two types of early x-ray machines and explain how they differed in terms of electrical generation.
9. List the basic difference between the early gas and hot cathode x-ray tubes.
10. Give the name of the individual credited with inventing the "hot cathode" x-ray tube, the date of its development, and its significance in revolutionizing radiographic technique.
11. State the year that the manufacturing of glass plates ceased and explain why.
12. Give the names of the devices in the static and coil x-ray machines that regulated radiation output of early gas x-ray tubes.
13. List the maximum exposure times recommended for use with gas tubes and radiator type "hot cathode" x-ray tubes for radiography of the pelvis.
14. Give the titles of two books that were of great assistance in explaining the operation and manipulation of early x-ray equipment to early radiographers.
15. List the names of recommended reducing and fixing agents used to process early glass plate radiographs.
16. Define the post-processing methods known as "intensification" and "general reduction."
17. List three guidelines identified by M. K. Kassabian, designed to protect the patient from excessive radiation exposure and three guidelines to protect the early x-ray operator.

Introduction

Today's medical radiographers are extremely fortunate to have fantastic modern x-ray technology to work with compared to the technology available during the early days, (Figure 1-1). This incredible imaging technology evolved over a very long period of time from the fields of electronics, which may be documented as far back as 600 B.C. (Hodges, 1939), and "photography, which originated in the first part of the 1700s" (Carroll, 1985).

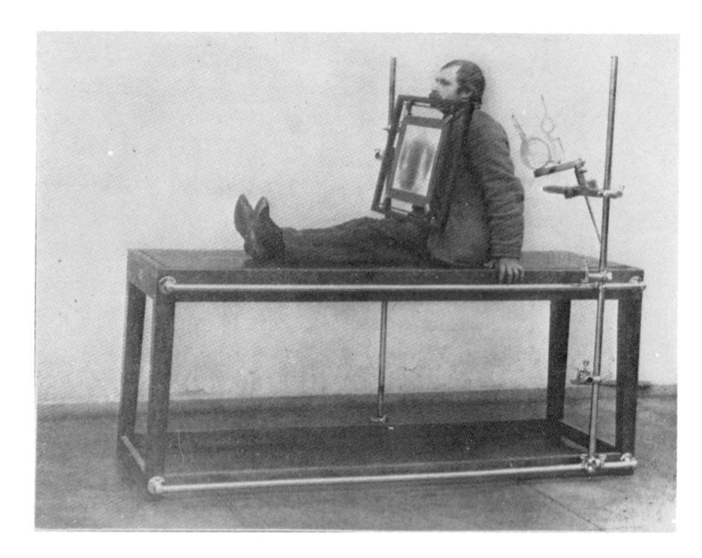

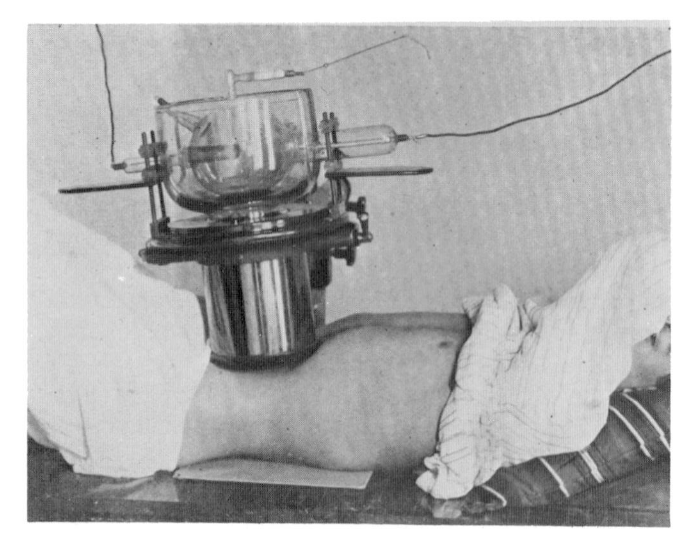

Figure 1-1. State of the art x-ray technology during the early days.

(See Appendix for a comprehensive listing of significant events leading up to and beyond the discovery of the x-rays.)

This chapter focuses primarily on the period between the discovery of x-rays in 1895 and 1923, the time when almost every phase of diagnostic x-ray service reached the stage of development which is recognized by today's standards as modern (Hodges, 1939). Also known as the period of x-ray commercialization, it defines the time when radiology grew from infancy to adulthood in which professional guidelines were firmly established. A look at the early days of radiography is a logical prerequisite to building a proper foundation from which the study and appreciation of radiographic quality may begin.

Roentgen's Discovery

On March 27, 1845, Wilhelm Konrad Roentgen (Figure 1-2) was born to his parents, Frederick Roentgen and Charlotte Frowein, in Lennep, Rhenish Prussia (a town near Cologne, Germany). However, he grew up in Holland, where his family moved in 1848. Although he did not take kindly to the stereotyped course of studies prescribed by the public schools, he later demonstrated an exceptional aptitude in physics and mathematics, earning a

Figure 1-2. Wilhelm Konrad Roentgen, Ph.D., (March 27, 1845 to February 10, 1923), and the physics lab at the University of Wurzburg, Bavaria where the discovery of x-rays took place.

doctor of philosophy degree from the Polytechnicum in Zurich, Switzerland. His dissertation was on the physical properties of gases. As an assistant to August Kundt, then professor of physics at Zurich, Roentgen was persuaded to study experimental physics.

After following Kundt to the University of Strassburg in 1872, the acceptance of a professorship in physics at an agricultural college near Stuttgart in 1875, and then back to Strassburg as an associate professor in 1876, Roentgen became a professor of physics at Giessen in 1879, and in 1888 accepted the professorship at the University of Wurzburg, Bavaria, the university that had once refused him appointment to the faculty for lacking a high school diploma (Kraft, 1973).

The discovery of the X-ray ("X" indicating the mathematical symbol for unknown) by Dr. Roentgen during the afternoon of November 8, 1895 at the University of Wurzburg, Bavaria, was the culmination of a chain of important evolutionary events in photography and electronics in which knowledge and application advanced to the point where someone with the properly developed intellect, such as Roentgen, could make the discovery.

There were other scientists equally equipped intellectually to make the discovery in 1895 who were also studying the effects of cathode rays in vacuum tubes. For example, Leonard supposedly advanced the notion that an invisible light may be a possibility as he succeeded in passing cathode rays into the air through an aluminum window (Hodges, 1939). However, Roentgen is credited as the first to confirm this possibility and quickly deduced virtually every x-ray property known today. Sylvanus P. Thompson, physicist and founder of the British Roentgen Society eloquently describes the historic event this way.

> "On that day a light which, so far as human observation goes, never was on land or sea, was first observed. The observer, Professor Wilhelm Conrad Roentgen. The place, the Institute of Physics in the University of Wurzburg in Bavaria. What he saw with his own eyes, a faint flickering greenish illumination upon a bit of cardboard, painted over with a fluorescent chemical preparation. Upon the faintly luminous surface a line of dark shadow. All this in a carefully darkened room from which every known kind of ray had been scrupulously excluded. In that room a Crookes tube, stimulated internally by sparks from an induction coil, but carefully covered by a shield of black cardboard, impervious to every known kind of light, even the most intense. Yet in the darkness, expressly arranged as to allow the eye to watch for luminous phenomena, nothing visible until the hitherto unrecognized rays, emanating from the Crookes tube and penetrating the cardboard shield, fell upon the luminous screen, thus revealing their existence and making darkness visible.... From seeing the illumination by the invisible rays of a fluorescent screen, and the line of shadow across it, and of verifying the source of the rays to be the Crookes tube, was to the practiced investigator but the work of a few minutes. The invisible rays, for they were invisible save when they fell upon the chemically painted screen, were found to have penetration power hitherto unimagined. They penetrated cardboard, wood, and cloth with ease. They would even go through a thick plank, or book of 2000 pages, lighting up the screen placed on the other side. But metals such as copper, iron, lead, silver, and gold were less penetrable, the densest of them being practically opaque. Strangest of all, while flesh was very transparent, bones were fairly opaque. And so the discoverer, interposing his hands between the source of the rays and his bit of luminescent cardboard, *saw* the bones of his living hand projected in the silhouette upon the screen. The great discovery was made (Thompson, S.P., 1897)."

Within the above scenario depicting Roentgen's discovery of the X-Ray can be found the prerequisites for radiographic imaging which qualified Roentgen as the original radiographer. The first prerequisite is that a source of x-rays must be made available. Roentgen isolated his source of x-rays from a Crookes-Hittorf tube that was conducting high-voltage electricity (Figure 1-3). The second prerequisite is that an object must be available from which distinctly recognizable shadows can be made. Roentgen supposedly used many objects including metal weights in a wooden box and his hands. The third prerequisite is that a recording medium must be available upon which the action of the x-rays can be observed as they traverse objects.

Roentgen first observed the presence of x-rays on a piece of cardboard upon which was painted a fluorescent chemical preparation of barium platinocyanide. Images obtained from this type of recording medium would be considered temporary since the observed image of an object was only available while the x-rays passed through it. Later Roentgen would use light-sensitive gelatine on glass plates (ordinary photographic plates) to permanently record images of objects as tangible evidence of his observations for others in the scientific community to verify.

Roentgen, as the original radiographer, persuaded Mrs. Roentgen to place her hand on a holder loaded with a photographic plate. Then for 15 minutes, Roentgen directed on it the emanation from one of his tubes (Glasser, 1945). On development, the

Figure 1-3. Crookes-Hittorf tube used by Dr. Roentgen for the production of x-rays.

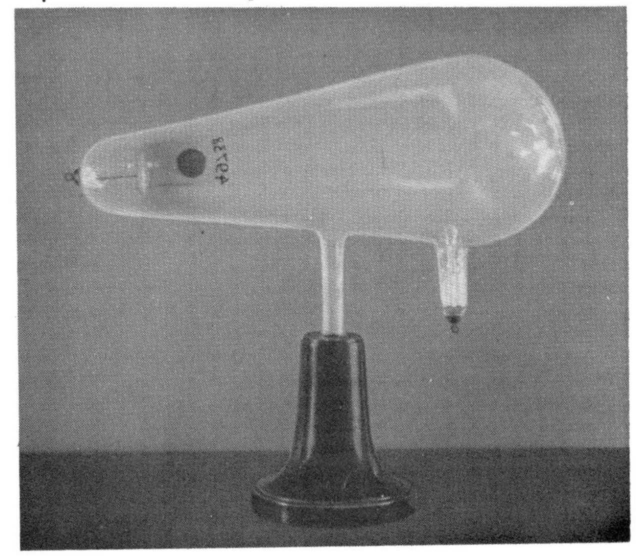

bones of her hands appeared; thus qualifying Mrs. Roentgen as the first radiography patient (Figure 1-4).

One can only imagine the incredible rush of excitement Dr. Roentgen must have felt as he immediately began to deduce the medical and industrial implications of his discovery. For the first time, the internal structures of the human body and inanimate objects could be viewed without resorting to surgery or disassembly. Roentgen's use of photographic plates to make permanent objective records of his results is the same basic radiographic imaging principle used today in conventional radiography that provides the physicians, engineers, scientists, and historians with the objective records of the status of a patient's health, the soundness of a manufactured item, and an understanding of historical and archeological findings. Roentgen's discovery is ranked as one of the greatest to benefit humankind. For his efforts he would receive many awards, one of which was the notable Nobel Prize in Physics given in 1901 (Sante, 1962). Roentgen followed the development of radiology, a new science, from its humble beginning to its great position as a tremendous medical diagnostic aid.

Dr. and Mrs. Roentgen did not have children of their own, but adopted Mrs. Roentgen's niece, who became their legal heir. The inflation that followed World War I wiped out their savings. The closing years of Dr. Roentgen's life were overshadowed by the loss of his wife, Berta, who died on October 31, 1919. Dr. Roentgen died at age 78 after a brief illness on February 10, 1923. He had cancer of the rectum which was diagnosed in its terminal stage (del Regato, 1985).

The Period of X-Ray Commercialization: 1895-1923

Immediately after Roentgen's great discovery was publicized, scientists and nonscientists alike began to reproduce his findings on self-constructed x-ray equipment or from equipment bought and assembled from new commercial x-ray equipment manufacturers. "At the forefront of the commercial x-ray movement were the electrologists. Electrology was a legitimate pseudo-science around 1900 although some of its practitioners were not. Electrologists had the technical knowledge and machinery to produce x-rays and an industry catered to their needs. They provided the first formal roentgen teaching, supported the first American X-Ray Journal in 1897 and participated in the first Roentgen Society of the United States. Around this time, an epic battle was fought to divorce roentgenology from electrology in which bitter words and hard feelings remained until the death of the respective participants," (Grigg, 1965).

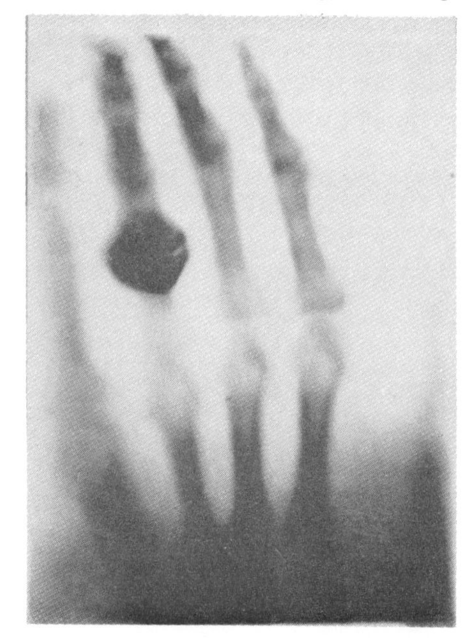

Figure 1-4. One of the first radiographic images produced by Dr. Roentgen depicting the structural details of Mrs. Roentgen's hand.

As commercial x-ray laboratories flourished across America and throughout the world, more and more people had access to x-ray services of physicians and commercial x-ray laboratory operators. One well-known commercial x-ray operator and roentgen pioneer, was Wolfram C. Fuchs (the father of Arthur W. Fuchs, a leader in the radiography profession between the 1930s and 1960s). "He established the first commercial x-ray laboratory in Chicago in May, 1896. A few years later, his business advertisement indicated that his x-ray laboratory was not only recognized by leading physicians and surgeons of Chicago, but that over 3,500 skiagraphs had been processed during the previous three-year period. W. C. Fuchs operated his successful and highly regarded laboratory until the effects of severe roentgen dermatitis forced his retirement in 1905 and his death in 1907," (Callear, 1970).

The term "skiagraphy" was primarily used to describe what is known today as "radiography." In the early days of radiographic imaging, skiagraphy was used to define the art of photographing shadows on sensitive glass plates by means of transmitted light. Skiagraphy was the preferred term since it was easy to pronounce, more general in meaning, and more euphonious. Other terms used included, Roentography, Shadowgraphy, Ixography, Electrography, Skotography, Kathography, Fluorography, Actinography, Radiography, Diagraphy, Skiography, Psyknoscopy, New Photography, and Electro-Skiagraphy. The main difference between ordinary photographs and skiagraphs was that photographs were perceived as *reflected* images of an object while skiagraphs were actual *transmitted* silhouettes (Kassabian, 1910).

The Birth of the Radiography Profession

The medico-legal aspects concerning "Roentgen-Ray Burns" (radiodermatitis) was probably the single most important factor that caused the medical professional to begin the process of gaining control over the commercialization of the x-ray. Almost immediately after Roentgen's discovery did lawsuits appear against physicians and commercial x-ray operators from patients who received roentgen ray burns. As more and more juries began to award plaintiffs verdicts in the thousands of dollars, physicians began to seriously question the scope of their responsibility with respect to the use of x-rays in diagnosis.

M. Hennecart at the Roentgen Ray Congress in Berlin in 1905, advocated legislation restricting to physicians the use of x-rays on human beings since it would be difficult to punish laymen in the event of avoidable injury. A resolution was adopted calling upon physicians to employ only medical men in x-ray work. A committee appointed by the Paris Academie de Medecine in 1906 concluded that since the medical use of Roentgen Rays may lead to serious accidents, and that certain practices may prove a social danger, only qualified physicians, health officers, or licensed dentists should be restricted in the use of x-rays (Kassabian, 1910).

When America entered World War I in 1917, there was a great need to train large numbers of men as military roentgenologists and assistants known as x-ray manipulators for service in the Army at home and abroad (Lauer, 1985). The use of the x-ray in the Spanish-American War of 1898 had proven to be a tremendous help to surgeons for preserving the asepticity of bullet wounds by doing away in many cases with the necessity for immediate exploration especially for lodged bullets and shrapnel fragments (Borden, 1900). As a consequence of meeting the roentgenology manpower requirements for the United States Army Medical Department during World War I, hundreds of physicians with no previous knowledge and skill in roentgenology were trained. Twice as many enlisted men were formally trained as Army x-ray manipulators (Lauer, 1985) (Figure 1-5).

When World War I ended and the Army began massive demobilization, hundreds of formerly trained Army roentgenologists and their x-ray manipulators returned to civilian status and continued to carve out a living using their newly acquired knowledge and skills of roentgenology. Such a sudden surge of manpower into the civilian commercial x-ray marketplace exacerbated and escalated the deterioration of medical ethics where the prudent use of the x-ray was concerned.

The relatively young radiology profession, which originated around the turn of the century with the birth of the American Roentgen Ray Society in 1901 (Grigg, 1965), mobilized to meet the problem of x-ray commercialization that was quickly getting out of hand. They did so in 1920 by first organizing x-ray workers into a noncommercial association. Subsequent control over that membership came in 1923 through a certification agency that would be governed by radiologists "until 1936 when two radiographers were allowed to become board members" (Morton, 1952). All radiologists were encouraged by their profession to inform x-ray workers employed by them of the certification movement that was underway in 1923 (Christie, 1924).

Figure 1-5. X-ray manipulators of the United States Army during World War I with an Army field x-ray apparatus that could be disassembled for transport within three minutes.

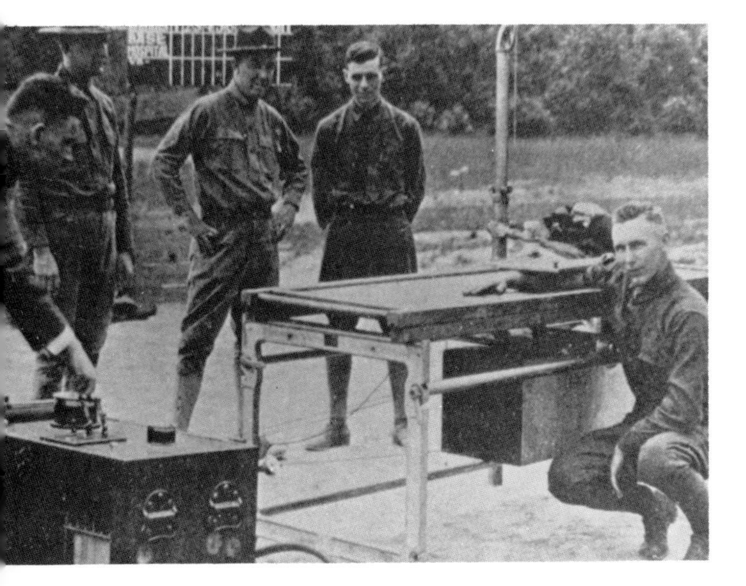

Figure 1-6. Ed C. Jerman (right) and sitting (below) for an organization photograph with charter members of the American Association of Radiological Technicians, 1920.

The person selected by leading radiologists to catalyze the certification movement through the development of a noncommercial association of x-ray workers was Ed C. Jerman (Figure 1-6), a man with twenty years prior experience as a former x-ray salesman, serviceman, counselor (Milligan, 1976), and charter member of the American Roentgen Ray Society in 1901 (Grigg, 1965). After two business failures in 1916, Jerman was discouraged with his life's work and was ready for a new challenge. Cooperating with and supported by leading radiologists (Hoing, 1946), he ventured off to secure potential members. His efforts culminated in the successful organization of the American Association of Radiologic Technicians (today's ASRT), in 1920 (Holland, 1982).

Shortly thereafter, through the initiation of the American Radiological Society of North America and under the official recognition of the American Roentgen Ray Society, the American Registry of Radiologic Technicians (today's ARRT) was born in 1923 (McKnight, 1982). For his efforts, Jerman was aptly rewarded as the first-elected president of the AART and appointed the first examiner of the Registry, positions he held for several years (Hoing, 1946).

In return for certification by the Registry, and subsequent membership into the Association, x-ray workers had to adhere to a noncommercial ethical agreement requiring them to work only under medical supervision, never to diagnose, and never to work independently in a private office, hospital, or institutional laboratory (Christie, 1924). Setting the perfect example for other certified radiographers to follow would be the Registry's first certified member, sister M. Beatrice Merrigan of St. Louis (McKnight, 1982). The control that radiologists sought over x-ray commercialization was further underscored through the passage of the Classification Act in 1923 which classified x-ray technicians as subprofessonals who worked under the supervision of the radiologist (Gaynor, 1951). Thus, the period of x-ray commercialization officially came to an end in 1923.

Early X-Ray Machinery

Around the turn of the century, the static and coil types of x-ray machines had been developed (Figure 1-7). The primary difference between these two types was based on their electrical generation. For example, in the static machine, electricity was generated by a motor unit carrying direct current to the x-ray tube. In the coil machine, primary electrical current of low potential was supplied from primary batteries, storage batteries, or dynamos and stepped up to a higher potential in the secondary portion of a Ruhmkorff induction coil.

The coil x-ray machines used by the United States Army Medical Department during the Spanish-American War of 1898 were limited to 10 type W Edison-Lelande cells in direct series which had a life of 600 ampere hours (200 working hours) at an operational cost then of 11 cents per hour. Although both static and coil-type x-ray machines of this period weighed about 500 pounds each, both produced equal power and working efficiency. However, the coil machines were less bulky, contained less glass, and were easily transported compared to the static machines. A popular manufacturer of the static machine was Otis Clapp & Son, or Providence, Rhode Island. Popular coil x-ray machine manufacturers of this period included the Edison Manufacturing Company, the Fessended Company, and the General Electric Company (Borden, 1900).

Figure 1-8. Diagram of Snook "hydrogen gas" x-ray tube.

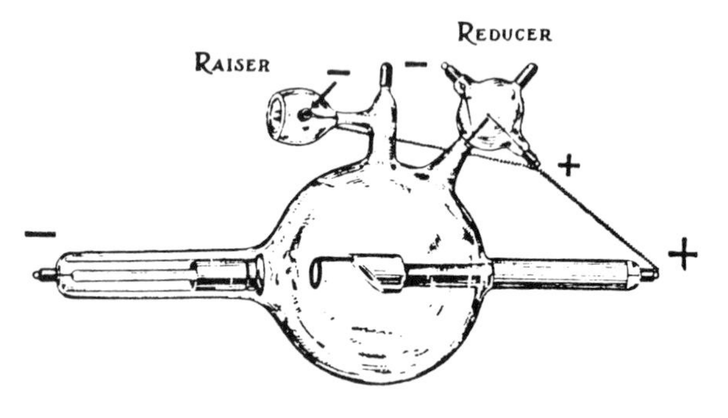

Early X-Ray Tubes

Early x-ray tubes were gaseous and varied with respect to their vacuum pressures (Figure 1-8). Therefore, they were divided into two categories, low or high. Such were the conditions of tubes received from the manufacturers. A low tube indicated one that readily produced low-penetrating radiation using a small amount of current. Low tubes were considered fit only for radiographing thinner body parts. High x-ray tubes produced greater penetrating radiation and were considered necessary for thicker body parts. Kassabian indicated that with a hard tube, short exposure, and proper development, soft tissue differentiation was attainable with the advantage being a reduction in motion, due to short exposure,

Figure 1-7. Two commercial x-ray machines used by the United States Army in the Spanish-American War (1898): Static type (left) and Coil type with batteries on (right).

and less radiation absorbed by the patient (Kassabian, 1910).

All tubes were recognized as having a life. A low tube by gradual use would become so resistant to current that the internal gas inside would no longer illuminate. (If the discharge of the tube was passing through the tube in the correct direction, the body of the tube would be filled with a uniform yellow-green fluorescence and that part of the tube above the flat platinum anode remained in shadow.) When this occurred, the vacuum could be lowered by subjecting the tube to heat by baking it in a hot-air oven, or carefully heating it over a Bunsen burner or alcohol lamp. If the tube did not respond to these methods, it could be returned to the manufacturer for reexhaustion (Borden, 1900).

Care was taken to not allow first time used, focused x-ray tubes to become red hot about their anodes since this would result in blackening of the inside of the x-ray tube due to particles of platinum being thrown off from the anode. This would result in a higher vacuum which decreased tube life. In this case, an alcohol lamp could be applied to the surface of the tube to warm it and decrease the vacuum. Radiographers were encouraged to always keep an eye on the x-ray tube while it was in use, since the vacuum was always in a state of fluctuation. Immediate action to reduce heat build-up in the gas tube required the use of a rheostat switch which controlled current to the machine. Considerable care was taken to make sure that x-ray tubes were always free from dust by frequent cleaning to avoid electrification of the glass envelope (Borden, 1900).

A ready method for determining the low or high condition of early x-ray tubes was to see whether or not a coin or cuff button held against the forearm could easily be seen through the radius. (Little did many of the early pioneers realize that they were being excessively exposed to hazardous x-rays by employing such a method.) A low tube could be raised to a high status by reversing the current for only a brief period. In the coil machine this was easily done by reversing the switch lever. In the static machine, closure of the current and reversal of the x-ray tube by the radiographer was required. High tubes required larger amounts of current for the production of greater penetrating x-rays. The test for determining high-tube status was to see with a fluoroscope the shadow of a watch through a man's skull (Borden, 1900).

Coolidge X-Ray Tubes

In 1912, Dr. W. D. Coolidge invented the "hot cathode" x-ray tube (Figure 1-9) which bears his name. This is classified as one of the most significant inventions in x-ray production since Roentgen's famous discovery. Together with other important technical advances, the interrupterless transformer of Snook (1907), the antiscatter diaphragm of Bucky (1911) and Potter (1916), and Eastman Kodak's double-coated film (1918) it permitted duplication of any given radiographic exposure simply by following numerical data. In 1917, Dr. Coolidge improved upon his hot-cathode tube by bringing out the Radiator-Type Coolidge X-Ray Tube from the General Research Laboratory (Figure 1-10) (Grigg, 1965). The radiator disks added to the x-ray tube anode section were for cooling purposes. These were 10 mA tubes

Figure 1-9. Diagram of Coolidge "hot cathode" x-ray tube introduced in 1912 with cathode insert close-up. Key to numbers on tube: 1. cathode terminal, 2. electron focusing cone, 3. solid tungsten target, 4. molybdenum supporting rod, 5. anode terminal. Key to letters on cathode insert: A. spiral filament tungsten wire, B. molybdenum "electron focusing" cylinder.

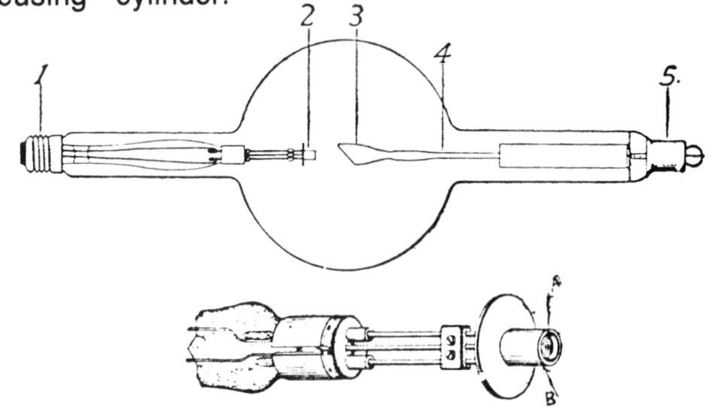

Figure 1-10. Diagram of radiator-type Coolidge "hot cathode" x-ray tube introduced in 1917.

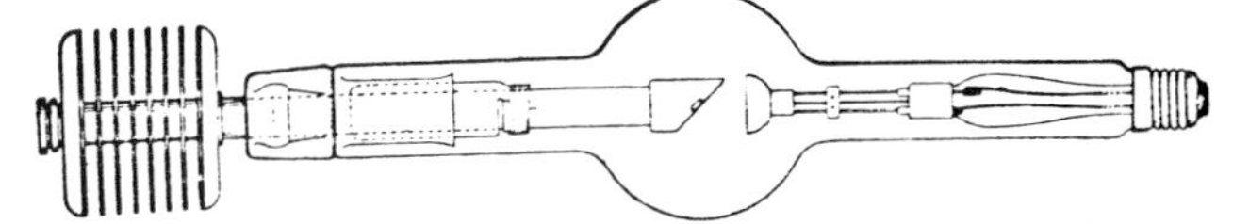

designed for use at a 5-inch spark gap during continuous fluoroscopy (U.S. Army X-Ray Manual, 1918).

The greatest difficulty in operating gas x-ray tubes was their irregular supply of electrons and the impossibility of control of their development from cold cathodes. In the Coolidge tube the cathode is a spiral filament of tungsten wire heated to a high temperature by a current from an insulated storage battery or by special transformer. The form of electrostatic field needed for focusing the electron stream is fixed by a small molybdenum cylinder within which the cathode is placed. The target was usually a solid piece of wrought tungsten mounted on a molybdenum rod, around which (in the radiator type tube) collars are placed to distribute the heat conducted from the target. Great care was taken in creating a very high vacuum nearly to the melting point of the internal metal parts.

There were several advantages to using the Coolidge tube. As long as the focal spot did not overheat, inverse current was suppressed. Operational voltage was about 100 kilovolts. It would operate without a rectifier if the focal spot was at a temperature below that at which it gives off an appreciable number of electrons (U.S. Army X-Ray Manual, 1918).

Early Radiography Recording Mediums

Roentgen remarked in his first communication concerning his discovery that he recorded his observations of x-rays on the fluorescent screen by means of photography and indicated that photographic films as well as glass plates may be used equally for this purpose. Although at the time of Roentgen's discovery the action of x-rays on silver salts in gelatin was not understood, by 1897 the relationship between this action became established. The early photographic glass plates were not very sensitive to x-rays. This prompted film manufacturers to begin producing plates that were sensitive to x-rays through thicker emulsion coats, which also reduced exposure times. The stoppage of glass being manufactured in Belgium during World War I (1914) (Bushong, 1980) occurred approximately one year after Kodak began to replace glass plates "By manufacturing a film having a cellulose nitrate base coated on one side with an emulsion more sensitive to x-rays than any previous available. In 1916, Kodak introduced a noncurling film base. In 1918, Eastman Dupli-Tized X-Ray Film was introduced as the first film designed for radiographic purposes that had a high-speed emulsion coated on both sides of the film base that permitted its use with double intensifying screens, which further reduced exposure" (Callear, 1970).

Because early x-ray glass plate construction required emulsions that were often unevenly poured, early radiographers were encouraged to make comparative tests before adopting a new batch of plates. This was done by placing four plates in light-tight envelopes and putting them side by side in the form of a square under the Crookes tube. On each plate was placed a circular disk of tin and in its center was placed a coin. The plates were then exposed for three minutes and then developed in the same developer tray and solution. The finished plate with the greatest density through the tin disk was determined to be the most sensitive and contrast was judged by the relative density between the shadow and the part of the negative outside it. Clearness of the negative and absence of chemical fog could be judged from the part of the negative beneath the place where the coin was placed. Since no x-rays passed through the coin, this part of the negative should have been perfectly clear (Borden, 1900).

Preservation of glass plates was stressed since they were frequently purchased in large quantities and shipped to distant areas with humid climates. Plates in separate light-tight envelopes did not keep well as the heat and moisture acted on the paper to cause deterioration of the emulsion. Plates lasted better when packed together in light-tight pasteboard boxes (Borden, 1900).

Since x-ray plates were extremely sensitive, they had to be stored in places that were absolutely free from smoke, gases, excessive light, etc. Plates stored in the x-ray laboratory were placed in wooden chests that were carefully lined with sheets of lead. Room temperature for storage purposes had to be consistent so as to avoid any possibility of moisture build-up that may cause mildew fogging. Packages of x-ray plates needed to be stored on edge to avoid undue weight pressure that could cause breakage. Early radiographers had to be very careful handling glass plates since the plates often had razor-sharp edges and small chip fractures (Borden, 1900). It would have heartened the pioneers of x-ray, who struggled in their cramped, ill-ventilated darkrooms with large clumsy glass plates, cutting their fingers on sharp edges and finding plates cracked in storage files in spite of all precautions, to know that by 1920, flexible x-ray film would have replaced glass (Hart, 1943).

Marking of glass plates was stressed since medico-legal cases indicated their importance. Negatives were marked with directional indicators such as the words "right" or "left" or abbreviations "R" and "L." Successive radiographs of the same part could use lead letters such as A B C D E etc. (Kassabian, 1910).

Early Radiographic Exposure Technique

The objective of the early radiographer using gas tubes was to manipulate the tube so as to secure maximum radiation output. In the static machine this was realized with the use of spark gaps and the velocity of rotation of the electrostatic glass plates. Spark gaps, also known as current interrupters, were metallic rods attached to and movable upon each sliding pole of the machine. High-vacuum tubes, resulting in greater penetration, worked best when wires were directly connected to the poles of the machine. Lower-vacuum tubes required increasing spark gaps. In the coil machine, a condenser (or in the case of coil machines with battery sets, a vibrator) was used to regulate maximum radiation output. After determining that maximum radiation output was secured for a particular tube, the part to be radiographed was scanned fluoroscopically in order to secure the best projection. Next, the plate was substituted to record the structural details of the anatomy (Borden, 1900).

Early radiographers became proficient through trial and error with sensing the efficiency of the radiation output with respect to the observable occurrences that took place before their very eyes. For example, with respect to the static machine, with the tube adjusted and the spark gap closed, the x-ray machine was started into rapid fluoroscopic action in which if the bones of the patient appeared gray and transparent, the tube was considered to be working at its best, in which no spark gap was required. On the other hand, to ascertain if fluorescence could be increased, the positive spark gap was separated and the results observed.

Then, the negative spark gap was separated and both gaps were adjusted to give the best radiation output from the tube. The distance was between a fraction of an inch greater. Although this procedure only took a few minutes, it was all a trial effort at determining the correct radiation output. Since on the static machine, the revolution of the electrostatic plates governed radiation output once the proper spark gaps were set, early radiographers were cautioned against increasing their revolutions when a dull red color appeared on the entire anode area. This might result in x-ray tube destruction with no adequate compensation in radiation output (Borden).

The length of radiation exposure depended upon the amount of radiation from the tube, the distance of the tube from the plate, the sensitivity of the plate to x-rays, and the density of the part through which the x-rays had to penetrate. Experience was required on the part of the x-ray operator to estimate the proper exposure time required to produce a satisfactory image (Borden, 1900). When Ed C. Jerman asked one of the early radiographers what spark gap, milliamperage, and exposure time he used to produce lateral spine plates, the reply was, "I pay no attention to any of those things; I never look at the meter, measure the gap, or use the timer" (Jerman, 1928). Jerman wrote,

> "His manner of procedure was as follows: He started the motor and began moving the rheostat lever backward and forward, watching the action of the tube and listening to the sound of the rectifyer and tube, and when he felt that

he had it set just right he proceeded to quickly place the plate and the patient in position. He then dropped his head between his shoulders and closed the X-ray switch, not even counting for time, and when "the hunch" told him, he pulled the switch, stopping the exposure. The plate was developed and the result [was some of the finest plates of the lateral spine ever seen]. He had worked with this equipment so long that he could tune it up as a violinist could tune his instrument" (Jerman, 1928).

When early x-ray gas tubes were operating at maximum radiation output, the shadow of an object through the skull of an adult during fluoroscopy was supposed to be clearly visible. At 10 inches from the body and using rapidly sensitive glass plates, the following maximum exposure times were recommended for use by early radiographers using gas tubes: (Borden, 1900)

Forearm and Hand . . . . 1 to 2 minutes
Shoulder and Chest . . . . 10 minutes
Knee . . . . 9 minutes
Hip, Head, Pelvis . . . . 20 minutes

Early radiographers recognized that penumbral effect (blurring) was minimal when there was a 65-degree angle between the film plane and the anode causing x-rays to fall most perpendicular to the film, which also resulted in minimal distortion. To determine if the x-rays fell perpendicular to the film plane, a metallic cylinder 3 to 4 inches long was placed on the x-ray plate and radiographed. Perpendicular rays cast a circular shadow, while oblique x-rays produced distorted elliptical shadows demonstrating nonsuperimposition of the two ends of the metallic cylinder. Recommended distance between the x-ray tube and film plane was 20 to 24 inches. The shortest possible distance that matched the field size of the part being radiographed was recommended whenever the shortest possible exposure time was desired. Although intensifying screens made of platino-cyanide of barium-shortened exposure time by a factor of 5 or 6 times compared to exposure by plates, they were not recommended for use since the finer details of the soft anatomic structures were sacrificed. The use of a beam-restricting device, like that depicted in Figure 1-11, reduced secondary radiation, but was not recommended highly since it also restricted field size which prevented obtaining comparison views of extremities (Kassabian, 1910).

Figure 1-11. Examples of flat (left) and tubular (right) x-ray beam restricting devices used by early radiographers for controlling secondary radiation. (Note that flat diaphragm on left is less effective at reducing secondary radiation compared to tubular diaphragm on the right.)

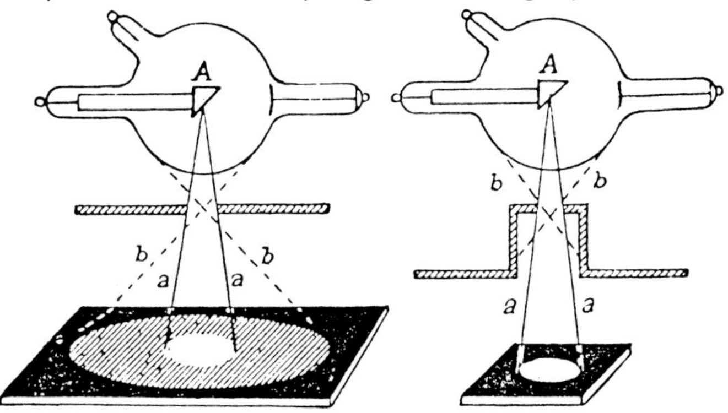

The way that the early "trial-and-error" radiographic exposure guidelines became increasingly valid and reliable was through a pattern of documentation of patient-exposure data. Figure 1-12, illustrates a form used by the Philadelphia Hospital Roentgen Ray Laboratory to record the patient's history and subsequent radiographic techniques employed (Kassabian, 1910).

With the advent of the "hot-cathode" Coolidge tube in 1912 and other devices such as the Snook transformer, Potter-Bucky antiscatter diaphragm in motion, calcium tungstate intensifying screens, and double-coated x-ray film, radiographic exposure techniques became more reliable. One of the first objective radiographic exposure technique charts devised was done by

Figure 1-12. Form used to document patient x-ray exposure data at the Philadelphia Hospital, Roentgen Ray Laboratory around 1910.

PHILADELPHIA HOSPITAL *No.*
RÖNTGEN RAY LABORATORY, RECORD OF DIAGNOSIS.

1 NAME | ADDRESS | Nativity / Occupation | Month | Day | 190
2 SEX Male / Female | Age / Single / Married | Color | Weight lbs. / Height | Department / Ward | Referred by M.D. / Address
3 Previous History
4 Date, Place, Duration, Character, etc., of Injury or Disease.
5 Part or Organ Involved
6–7 SYMPTOMS Physical and Clinical
8 Chemical and Microscopical Examinations
9 REMARKS

Technic Employed in Diagnosis.

I APPARATUS	II POSITIONS OF THE TUBE AND PATIENT	III QUALITIES OF THE RAYS, TIME OF EXPOSURE
A STATIC MACHINES: Varieties and Make; Revolving Plates; Number; Diameter inch / cm.; Rev. per Minute; Length of Spark-gap inch / cm.	Distance of Anode from Plate inch / cm.; Thickness of the Part inch / cm.	Current going to the Primary Coil Volts / Amp.; Secondary or Induced Current Milliampere; Parallel Spark-gap inch / cm.; No. of Benoist's Scale; Degree of Vacuum of Tube: Low (soft), Medium, High (hard)
B SOURCE OF CURRENTS: Accumulator Volts; Ampere-hour; Direct Current; Altern. Current; Transformer	POSITIONS OF PART: Antero-posterior; Lateral; Flexion; Extension; Dorsal Decubitus; Ventral Decubitus; Recumbent; Semi-recumbent; Sitting; Standing; With or without Bandage, Splint, Cast	Time of Exposure sec. / min.; Intensifying Screen; Variety of the Plate; Sizes / Nos. / Parts: X A, X B, X C; Negatives
C KIND OF COILS: Varieties; Length of Spark-gap inch / cm.		
D INTERRUPTERS: Varieties; No. of Interruptions; Per Minute; Mechanical; Mercury; Wehnelt; Caldwell; Simon	E CROOKES TUBE: Varieties; Non-Regulating; Self-Regulating; Osmo-Regulat.	Over or under-exposed or developed, Patient moved; No. of Prints; Duplicate

Diagnosis made from Skiagram or Fluoroscope or Stereo-Skiagraph

OPERATOR

AUTOPSY

Diagnosis made by Director .... *M.D.*
or Assistant .... *M.D.*

Lieutenant Colonel Shearer, Ph.D., (Figure 1-13) and printed in the physics section of the *United States Army X-Ray Manual* and its *Extract* published in 1918 at the end of World War I. These exposure times were supplied for Army x-ray manipulators working with Army x-ray field apparatus during World War I which used radiator-type Coolidge tubes. See Table 1-1.

All exposure times were assumed to be made using 5-inch spark gaps, 40 mA, and 20 inches from the anode target to the film plane. (Chests were taken at 28 inches.) The average man was assumed to be 150 pounds. Seed x-ray plates were used under direct exposure conditions. As anatomic part thickened, exposure times were recommended to increase proportionately. The Army being so adamant in the use of this objective radiographic exposure table, felt that any x-ray operator who couldn't improve on the results secured by adhering to any single exposure table was not fit for the work (*U.S. Army X-Ray Manual*, 1918).

TABLE 1-1. **Radiographic Exposure Table Recommended by United States Army During World War I**

Part Radiographed	Exposure Time in Seconds
Head AP	12
Head Lat	6
Neck	3
Shoulder	3½
Elbow	1½
Wrist	1
Kidney	3 to 5
Bladder	3 to 5
Hip Joint	5 to 7
Pelvis	5 to 7
Knee	2
Ankle	1½
Lumbar Spine	5 to 6
Teeth (Slow Film)	4
Teeth (Fast Film)	1½
Chest (at 28 inches)	2½ to 4

(Source: *United States Army X-Ray Manual*, Paul B. Hoeber, New York, 1918)

Dr. John Sanford Shearer was a physicist who also served as an officer in the United States Army during World War I. While in the Army he developed x-ray equipment, technical training literature, and taught Army medical officers and enlisted personnel x-ray physics and techniques. His authorship of the x-ray physics section of the *United States Army X-Ray Manual* published in 1918 (John Sanford Shearer, 1951) clearly and concisely presented considerable information concerning the relationship between exposure and density (the characteristic curve), definitions of radiographic density, contrast, and detail along with their affecting exposure factors.

The training of hundreds of Army x-ray manipulators by the Army at the Army Medical School, Fort Riley, and Camp Greenleaf, represented one of the first and largest formal training efforts for the training of medical radiographers in North America (Lauer, 1985). Prior to this, such training was either

Figure 1-13. Lieutenant Colonel John Sanford Shearer Ph.D., and cover page of U.S. Army X-ray Manual in which he wrote the section on x-ray physics and techniques.

done by apprenticeship in some hospital or commercial x-ray laboratory or by quick training sessions conducted by sales representatives of commercial x-ray equipment manufacturers. Knowing this, one has to wonder why Ed C. Jerman is given so much credit as the symbolic "Father of Modern X-Ray Technics" and the first to define the radiographic qualities (X Ray Techniquiz, 1965) of contrast, density, and detail instead of John Shearer since the *United States Army X-Ray Manual* was published almost ten years before Jerman published his book *Modern Radiographic Technic* in 1928.

Since the 1918 Army manual was published commercially by Paul B. Hoeber of New York, it was available to the community of American x-ray operators and was probably used extensively by a great many in some type of formal or informal capacity during the 1920s. Just how much this Army manual influenced the writings of Jerman, Sante, and others that would follow is not exactly clear. However, one can assume that many leaders in the field definitely knew of its existence.

The book, *Modern X-Ray Technique*, by Ed. C. Jerman, was considered one of the most helpful early publications to early radiographers and assisted in further standardizing the objective application of radiographic imaging techniques. Jerman's greatest technical contribution was in the reliable ordering of the four primary radiographic exposure factors: distance, time, milliampere-seconds, and kilovoltage. When taken into account with the use of grids and calcium tungstate intensifying screens, and optimal-developing conditions, diagnostic quality radiographs could be generated on a relatively consistent basis. (See Table 1-2 for some of Jerman's sample exposure techniques.) Accompanying Jerman's exposure techniques were photographs of the radiographic position to be performed and example diagnostic quality radiographs.

The unique aspect of Jerman's book was the style in which he wrote about the principles of radiographic technique. He delivered a concise and comprehensive discussion of the radiographic topics

TABLE 1-2. **Sample of Jerman's Recommended Radiographic Exposure Techniques.**

Region	Pos.	Time	Dist	mA	kVp	Time	Dist	mA	kVp
	With Screens					Without Screens			
ELBOW (Non Bucky)	AP	5	36″	10	54	10	25	10	57
	AP	2	36″	30	50	52	53	30	53
	AP	1/10	36″	100	66	3	36	100	60
SHOULDER (With Bucky)	AP	5	36″	10	62	5	25	10	73
	AP	2	36″	30	58	5	25	30	63
	AP	1/10	36″	100	74	3	36	100	74
CHEST (Non Bucky)	PA	1/10	48″	100	76				
	PA	1/5	72″	100	73				
	PA	1/2	72″	100	60				
	PA	3/10	40″	30	75				
	PA	1	72″	30	75				
KNEE (With Bucky)	AP	10	36″	10	60	10	25	10	64
	AP	5	36″	30	56	5	25	30	60
	AP	1/10	36″	100	72	3	36	100	70
PELVIS (With Bucky)	AP	6	25	10	88				
	AP	3	30	30	88				
	AP	1	25	100	75				
(Non Bucky)	AP	20	36	10	60				
	AP	2	36	10	88				
	AP	10	36	30	57				
	AP	3/4	36	30	88				

(Source: Jerman, Ed. C., *Modern X-Ray Technique*, Bruce Publishing Company St. Paul, Minneapolis, 1928)

followed by end-of-chapter questions and answers. No doubt Jerman, who served as the first Examiner for the American Registry of Radiological Technicians in 1923, knew that these questions and answers would be invaluable to early radiographers desiring to pass the Registry's radiography credentialing examination in an effort to join the certification movement.

Even after reprinting of Jerman's *Modern Radiographic Technique* ceased, the question-and-answer portion continued to be revised periodically and appeared under various titles until 1947. At that time, the question-and-answer portion was published under the title *Techniquiz*, a booklet that proved exceptionally popular. Subsequent revision of *Techniquiz* by the X-ray department of the General Electric Company continued to foster better understanding of the broader aspects of x-ray technique (X-Ray Techniquiz, 1965).

By the end of the decade of the 1920s and the beginning of the 1930s three more significant writing events transpired that would impact on desseminating the literature of radiographic exposure technique among the community of organized radiographers. LeRoy Sante, M.D., published a popular book similar to Jerman's entitled, *Manual of Roentgenological Technique* in 1928 (Sante, 1962). The American Society of X-Ray Technicians began publication of their journal entitled, *X-Ray Technician* in 1929 (Hoing, 1946). Kodak began production of a journal entitled, *Radiography and Clinical Photography* in 1930 (Arthur W. Fuchs Retires, 1961).

Early Radiographic Processing Methods

The technique for producing the radiographic image around the turn of the century did not differ much materially from that used in ordinary photographic work except for the electromagnetic energy used to expose film, length of exposure, and processing methods. Early radiographers with an understanding of the rudiments of photography were at a distinct advantage.

Developer solution used on x-ray plates consisted of four parts: reducer, preservative, accelerator, and restrainer. The best reducing agents were considered to be pyrogallic acid (also known as pyrogallol), metol, hydrochinone, ortol, eikonogen, and rodinal. Early radiographers were responsible for mixing their own chemistry. The formulas for producing pyrogallol, metol, and hydrochinone are as follows (Borden, 1900).

Pyrogallol

PART A

	Grams
Boiled Water	500
Potassium Bromide	1
Sulphite of Soda	125

Part B

	Grams
Boiled Water	500
Sodium Carbonate	125

Metol and Hydrochinone

	Grams
Boiled Water	500
Sodium Sulphite	50
Potassium Bromide	1
Metol	2
Hydrochinone	6
Sodium Carbonate (crystallized)	50

Since it was known that all reducing agents experienced easy oxidation, sodium sulphite was added as a preservative. The action of the developer was regulated by the addition of an accelerator or a restrainer. Slow-acting developing solutions required the addition of the accelerator known as carbonate of sodium or potassium while fast-acting developers required the restrainer potassium bromide. When greater contrast was desired, hydrochinone along with a caustic alkali as an accelerator, and some potassium bromide as a restrainer, gave best results; ortol gave good results and had the advantage of being ready mixed only by applying water. The following steps were used by early radiographers in the development of x-ray plates (Kassabian, 1900):

Developing Steps

1. Immerse plates in a tray of cool water before placement in a developer tank or tray.
2. After tank immersion, move plates up and down quickly to prevent air-bells or bubbles from forming on the plate surface.
3. After 5 or 10 minutes in the tank, lift plates out of tank and reverse position placing that end of the plate which was at the

top of the tank to the bottom. (This prevented streaking.)
4. Rock or shake the tank every 5 minutes during development to prevent streaks or spots.

Developing times depended on the thickness of the anatomical part that had been radiographed, assuming the developer solution was fresh and at the recommended temperature of 65 or 60 degrees F. For example, development of radiographic images of the abdomen took considerably longer than radiographs of the hands or feet. If the image appeared within a couple of minutes, this was an indication that the x-ray plate had been underexposed. In this case accelerator was added to the developing solution. If greater image details were desired, plates were developed in tanks without the addition of accelerator between 40 or 60 minutes (Kassabian, 1900).

After development, the plates were washed in running water before being placed in an acid chrome fixing bath. This fixing solution was recommended over plain "hypo" fixing solution since it did not discolor and kept in storage longer. The purpose of the fixing solution was to remove the developer stains from the negative plates and harden the plate emulsion. Filtration of fixer was recommended before use. The following formula was used for producing acid chrome fixer: (Kassabian, 1910)

Acid Chrome Fixer

Water	100 ounces
Sulphuric Acid	3 ounces
Sulphite of Soda	4 ounces
When dissolved add Hypsulphite of Soda	2 pounds

(Dissolve and add Chrome Alum 1 to 2 ounces previously dissolved in 20 ounces of water. Next add water to make a total of 160 ounces.)

The plates were recommended to remain in the fixing solution 3 to 5 minutes to insure hardening of the radiographic image. Another 5 minutes was recommended to allow for a thorough fixing for x-ray plates having thicker double-coated emulsions (Kassabian, 1910).

After fixing, washing was immediately done for a period of one hour. If running water was not available, the plate was placed in a flat dish and constantly rocked for 5 to 10 minutes before water changes. In hospitals and large x-ray laboratories it was useful to employ two large wooden boxes that acted as tanks with different-sized compartments to handle six x-ray plate sizes. One of these tanks would contain sufficient fixing solution for six months and the other tank would be used for the final washing (Figure 1-14). After fixing and washing, the plate was hardened further by placing it in a solution of 4 ounces of water and one ounce of formaldehyde for 5 to 10 minutes (Kassabian, 1900).

After the final wash and hardening, plates were dried in a shaded room of moderate temperature and sufficient air ventilation. Drying plates in the sun was not recommended since this would increase density and soften the emulsion. A fast-drying technique was to place the plate in a bath of alcohol for 5 to 10 minutes after washing thoroughly and then place it before an electric fan (Kassabian, 1900).

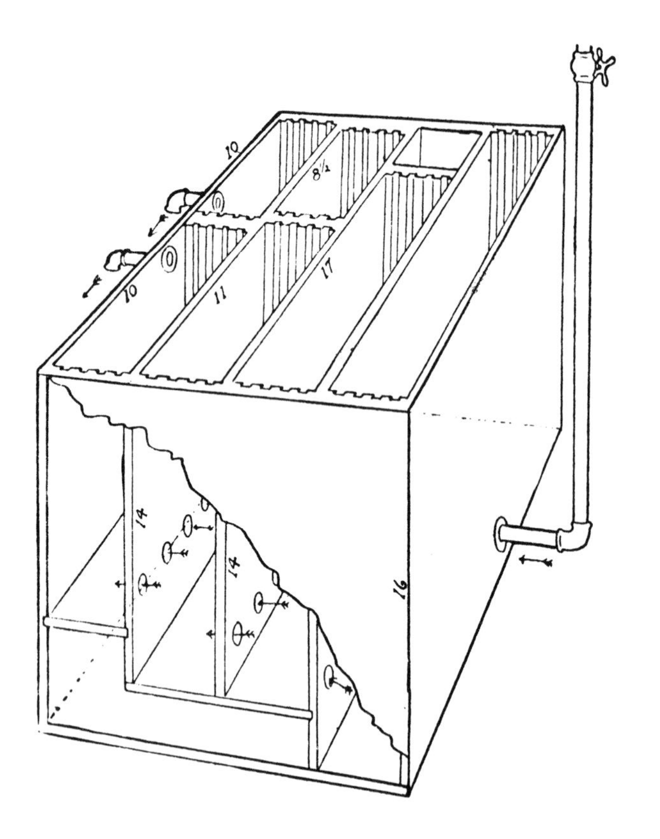

Figure 1-14. Early fixing and washing tank construction. (Note that wash water enters at bottom and circulates to two overflowing discharge pipes.)

Radiographic plates that were underexposed could be darkened slightly after the normal drying procedure was completed through a process called "intensification." The solution for intensification involved 13.3 grams of mercuric bichloride, 8.0 grams of potassium bromide, and 6.5 ounces of water. After the plate was immersed in this solution and became uniformly bleached (the longer it was bleached, the more dense became the negative), it was thoroughly rinsed and washed under running water for half an hour and then blackened using one of the following solutions (Kassabian, 1910):

Blackening Solutions Used in Intensification

Sodium Sulphite	30 grams
Water	120 cc
OR	
Ammonia	1 cc
Water	30 cc

After blackening, the plate was washed and dried in the usual manner previously described. Any yellowish coloration remaining was an indication of insufficient washing.

Overexposed plates could likewise be lightened through a process called "general reduction." The solution used for this purpose had two parts (Kassabian, 1910):

PART A

Water ....500 cc

Hyposulfate of Soda ....30 grams

PART B

Water ....500 cc

Potassium Ferricyanide ....30 grams

After mixing 8 parts of solution A with 1 part of solution B in subdued light, the negative was placed in this solution directly after fixing. If a dry negative was to be reduced, it had to be soaked in water first for at least half an hour. After sufficient reduction was achieved, the negative was washed and dried in the usual manner (Kassabian, 1910).

Early Radiation Protection Guidelines

Producing the early radiographic image was not only a tedious job, but one of considerable risk with respect to excessive radiation exposure to the patient and the operator. Almost immediately after Roentgen's discovery were the effects of excessive radiation exposure reported upon. However, a general lack of understanding caused many to take unnecessary risks. While patients received x-ray burns, many operators contracted radiodermatitis which would later metastasize and eventually result in the victim's death.

An early lawsuit concerning x-ray burns suffered by a patient was brought against Dr. Otto Smith and Professor W. C. Fuchs of Chicago for $25,000.00.

> "The plaintiff, aged 37, broke his right ankle as the result of an accident on September 2, 1895. He was able to attend his business on May 1, 1896, and was practically as well as ever. He only suffered from slight stiffness and occasional swelling in the ankle. On September 19, 1896, x-ray photographs were made, each sitting occupying from 35 to 40 minutes, the tube being placed 5 or 6 inches from the ankle. While under the exposure, the patient complained of sharp, tingling pains. Three days after, a slight redness appeared between the big toe and the adjoining one, which in three weeks had spread over almost the entire dorsum of the foot, later forming a blister. An intensely painful ulcer formed, for which condition amputation of the foot was performed. The jury awarded the plaintiff a verdict for $10,000.00." (Kassabian, 1910)

During the Spanish-American War of 1898, two radiation-burn cases were reported; one being from a coil x-ray machine and the other from a static machine. The x-ray burn from the coil machine involved Thomas McKenna, a discharged soldier who had been treated for a gunshot wound to the upper third of the right humerus during the Santiago Campaign. In an attempt to radiograph the shoulder on December 5, 1898, to ascertain the condition of the bone, a 20-minute exposure was made using a "low" tube positioned 10 inches from the shoulder. The results were so poor that second and third attempts were made on successive days, which also resulted in unsatisfactory radiographic images. Six days after the last exposure, a slight redness of the skin about the shoulder area was noticed. Two days later, the erythematous condition increased along with the appearance of small blebs. These blebs broke and small ulcers formed which gradually spread. The tissue necrosis deepened, extended, and was accompanied by marked pain and hyperaesthesia. The inflammatory action continued until the burn covered nearly the whole right breast (Figure 1-15). Treatment producing the greatest results included lead and opium lotion. The burn showed no signs of healing after 4 months. The healing process was very slow and the burn was not entirely healed until eleven months from its first appearance (Borden, 1900).

Figure 1-15. X-Ray burn of Thomas McKenna, (1898).

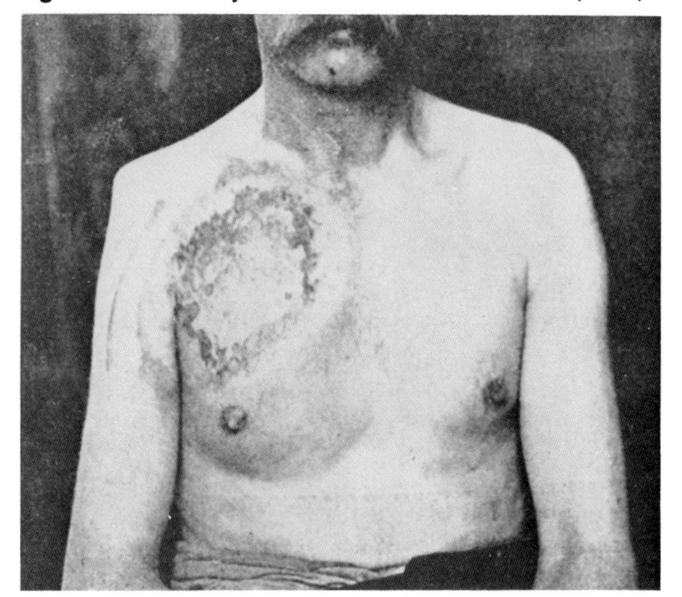

The burn from the static machine involved a Private Walter C. Booth who in 1899 had been admitted to the hospital for pyelitis and hydronephrosis of the left kidney supposedly due to the presence of a calculus in the renal pelvis. A radiographic attempt to determine the presence of the renal calculus resulted in 3 exposure attempts lasting 25 minutes each. No calculus was located and five days after the last exposure, an erythematous spot appeared on the left side of the abdomen which became quite pronounced in color. There was hyperesthesia of the part, but no ulceration. The irritation disappeared in about 10 days leaving no after effects (Borden, 1900).

Between 1899 and 1910, a period of eleven years, Dr. Mirhan Krikor Kassabian (roentgen pioneer) documented the deterioration of his hands as a result of contracting radiodermatitis early in his career. Several fingers had to be removed by amputation. The striking photographs in Figure 1-16, provide graphic representation of the destructive power of x-rays on human tissue when radiation protective measures were not fully understood or applied during long-exposure periods, including fluoroscopy. Obviously, radiodermatitis had a significant debilitating effect which reduced the quality of life of its victims. On July 14, 1910, Dr. Kassabian died in Philadelphia following metastatic malignancy of the axilla (Mihran Krikor Kassabian, 1949). Dr. Kassabian's final advice to the x-ray operator was that, "It is far bet-

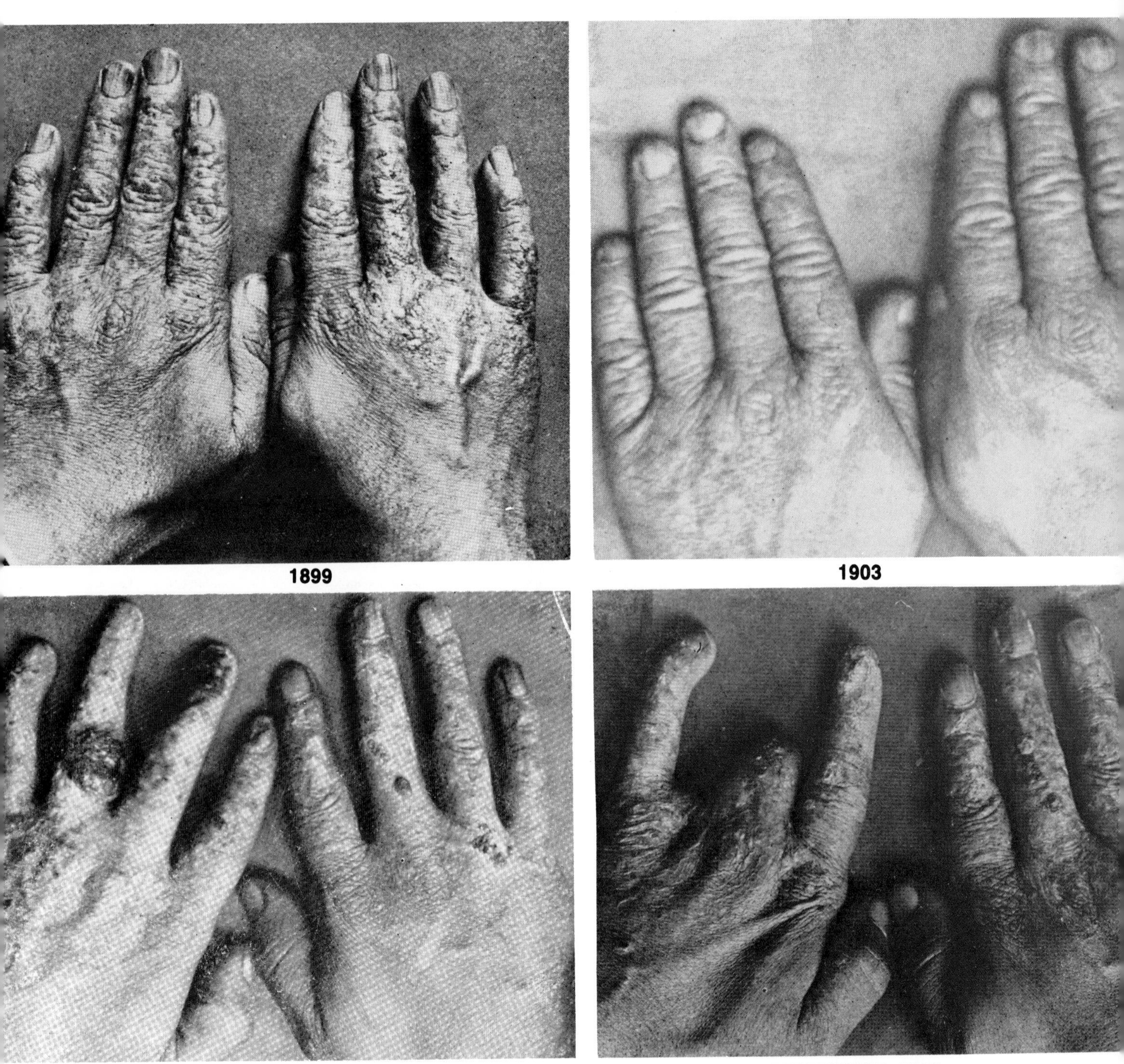

Figure 1-16. Photographic documentation from 1899 to 1909 of the deterioration of Dr. Kassabian's hands due to radiodermatitis induced by excessive exposure of his hands to x-rays. (1899 and 1903 photographs are in proper sequence.)

ter...for the operator to be out of the room and dispense with protective devices'' when patients are undergoing radiation exposure (Kassabian, 1910).

Guidelines identified by Kassabian to protect the patient from excessive radiation in the early days of radiography included:

1. Patients undergoing x-ray examination should only be given small interrupted doses.

2. High (gas) tubes produce less radiodermatitis then low (gas) tubes.

3. The more distance from the tube, the less will be the danger of burning the patient.

4. Surrounding healthy parts of the body should be screened from x-rays by thin sheets of lead.

5. Filtering off the soft and unnecessary rays should incorporate the use of a piece of leather or aluminum (grounded) applied over the part to be radiated.

6. Relief of symptoms of acute radiodermatitis should incorporate the use of the compound zinc powder with 10 percent icythyol.

Guidelines identified for the protection of the x-ray operator included:

1. The operator should never use his hands to test the intensity of the x-rays.

2. While the patient is undergoing treatment or examination,

the operator should be in a communicating room, behind a lead screen, or in a sentry box, where he can observe the fluorescence of the tube from a mirror suspended at a convenient angle from the ceiling (Figure 1-17).

3. The best place for lead protection is behind the anode.
4. The absence of the operator from the room is advisable since its atmosphere becomes ionized by x-rays.
5. To discover the presence or absence of the x-rays requires the use of an electroscope or photographic plate. Whenever x-rays are detected, that region of the room is unsafe for the operator.
6. Opaque rubber gloves, aprons to prevent injury to the testicles, and lead glass spectacles should be worn for added protection.

SUMMARY

The discovery of the x-ray on November 8, 1895 by Dr. Roentgen in the afternoon at the University of Wurzburg, Bavaria, was followed by a long and intense period of x-ray commercialization between 1895 and 1923. One of the early roentgen pioneers credited with the establishment of the first commercial x-ray laboratory in Chicago in 1895 was Wolfram C. Fuchs. Skiagraphy, shadowgraphy, and actinography are examples of early terms that describe what today is known as radiography. Ed C. Jerman in 1920 is credited with organizing the American Association of Radiological Technicians, AART, in 1920 with the assistance of organized radiology so that control over x-ray commercialization could be achieved certifying x-ray operators. Membership into AART was dependent upon certification by the American Registry of Radiological Technicians, an organization initiated in 1923 by the American Radiological Society of North America and officially recognized by the American Roentgen Ray Society. Ed Jerman served as the first president of AART and first examiner of ARRT, positions he held for several years. The first to be certified by ARRT was Sister Beatrice Merrigan of St. Louis.

The two types of early x-ray machines were named static and coil. Electricity was generated in the static machine by a motor-carrying direct current and in the coil machine, by batteries or dynamos. Radiation output was achieved in static machines through the velocity of the static plates and the use of spark gaps. In coil machines, radiation output was regulated primarily by a condenser (or a vibrator if battery sets were used). Early gas x-ray tubes were categorized as high (for greater x-ray penetration) or low (for less x-ray penetration). Gas x-ray tubes differed from hot cathode tubes in terms of controlling the electron supply to the anode. Dr. Coolidge, who invented the hot cathode x-ray tube in 1912, improved upon his invention when in 1917 he brought out the radiator-type Coolidge tube.

In 1914, the supply of Belgium glass was halted because of World War I. Around this time, Kodak began to manufacture cellulose nitrate base x-ray film. An example exposure time for radiographing a pelvis using a gas x-ray tube was 20 minutes. Using a radiator-type Coolidge tube, it was between 5 to 7 minutes. Examples of early reducing agents used in the development of x-ray glass plates included pyrogallo, metol, and hydrochionone. Acid chrome fixer was the preferred fixing solution since it did not discolor and kept in storage longer than regular hypo. A post-processing method that was used to slightly darken underexposed x-ray plates was called, "intensification." "General reduction" was the name given to the process of lightening overexposed x-ray plates once they were regularly processed. Patients and operators alike suffered from the effects of overexposure to x-rays during the early days of radiography. Mihran Krikor Kassabian, roentgen pioneer, died of cancer of the axilla as a result of severe radiodermatitis of his hands. Before he died, he provided several important guidelines concerning patient and operator protection from radiation exposure.

Figure 1-17. X-ray facility scheme recommended by Dr. Kassabian to protect x-ray operator from x-ray exposure (Note the indirect method of monitoring patient during exposure by way of a mirror.)

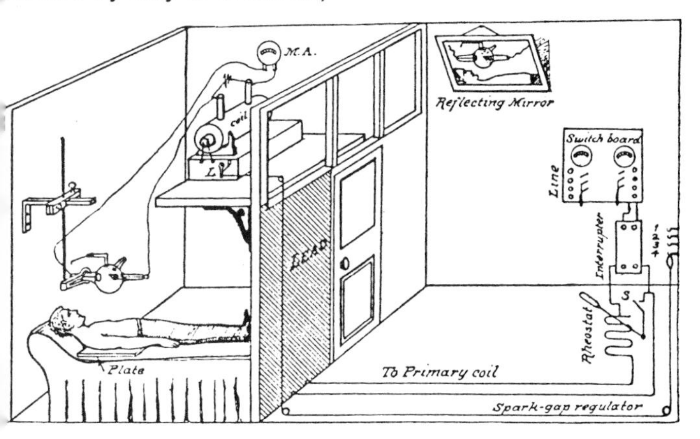

STUDY QUESTIONS

1. Roentgen discovered x-rays on ________ at the __________.
   a. November 8, 1895, University of Wurzburg
   b. November 8, 1898, University of Strassburg
   c. November 5, 1895, University of Wurzburg
   d. November 5, 1898, University of Strassburg
2. The period of x-ray commercialization encompassed the time between the discovery of x-rays and:
   a. The formation of AART
   b. The end of World War I
   c. The discovery of the rotating anode
   d. The formation of the ARRT
   e. The discovery of the hot cathode x-ray tube
3. The roentgen pioneer who established the first commercial x-ray laboratory in Chicago in 1896.
   a. Coolidge
   b. Jerman
   c. Fuchs
   d. Kassabian
   e. Borden
4. Which of the following was considered the preferred term used to describe what is commonly known today as radiography?
   a. Actinography
   b. Electrography
   c. Skiagraphy
   d. Shadowgraphy
   e. Diagraphy

5. Which of the following does not pertain to Ed C. Jerman?
   a. First examiner of the ARRT
   b. First president of AART
   c. Wrote the book, *Modern X-Ray Technic*
   d. Wrote the physics section for the United States Army X-Ray Manual
6. What year did almost every phase of diagnostic x-ray service reach the stage of development which is recognized by today's standards as modern?
   a. 1912 d. 1923
   b. 1918 e. 1928
   c. 1920
7. The hot cathode radiator type x-ray tube was invented by ______________ in ______________
   a. Jerman, 1910 d. Coolidge, 1912
   b. Shearer, 1912 e. General Electric, 1918
   c. Sante, 1915
8. The manufacturing of glass plates in America was halted in ______________ due to ______________
   a. 1898, Spanish-American War
   b. 1910, increased medico-legal cases
   c. 1913, the manufacturing of cellulose nitrate base x-ray film.
   d. 1914, World War I
   e. 1923, the Classification Act
9. A device that helped to secure maximum radiation output in coil type x-ray machines.
   a. spark gaps c. condenser
   b. current interrupters d. valve tube
10. Early gas x-ray tubes that had greater penetration capability were classified as ______________
    a. high d. pressurized
    b. low e. nonpressurized
    c. intense
11. A valid example of a recommended exposure time for radiographing a pelvis using an early gas x-ray tube that was considered to be in optimal working order.
    a. 5 minutes d. 30 minutes
    b. 15 minutes e. 45 minutes
    c. 20 minutes
12. Which of the following is not the name of an early reducing agent that was used in the development of glass plates.
    a. hypo d. hydrochinone
    b. pyrogallic acid e. rodinal
    c. metol
13. The name of the individual who wrote the physics and technic section for the United States Army X-Ray Manual that was published in 1918.
    a. Ed C. Jerman d. LeRoy Sante
    b. John Shearer e. W. D. Coolidge
    c. M. K. Kassabian
14. The name of the post-processing method used to lighten an overexposed x-ray glass plate radiograph.
    a. Subtraction c Reduction
    b. Addition d. Division

BIBLIOGRAPHY

1. "Arthur W. Fuchs Retires," *Medical Radiography and Photography*, 37(2):1961.
2. Borden, W. C., *The Use of the Roentgen Ray by the Medical Department of the United States Army in the War with Spain.* (1898.), Government Printing Office, Washington, 1900.
3. Bushong, Stewert C., *Radiologic Science for Technologists*, C. V. Mosby, St. Louis, 1980.
4. Callear, Thomas E., (ed), "Kodak and Radiography" *Medical Radiography and Photography*, 46(3):1970.
5. Carroll, Q. B., (ed), *Fuch's Principles of Radiographic Exposure, Processing, and Quality Control*, 3rd ed., Charles C. Thomas, Springfield, Ill., 1985.
6. Christie, Arthur C., (ed) "American Registry of Radiological Technicians," *The American Journal of Roentgenology and Radium Therapy*, XI(4):1924.
7. del Regato, Juan A., *Radiological Physicists*, American Institute of Physics, New York, 1985.
8. Gaynor, Valeria M., "General News: Commissions in the Armed Forces," *X-Ray Technician*, 22(4):1951.
9. Glasser, Otto, "Fifty Years of Roentgen Rays" Radiography and Clinical Photography, 21(3):1945.
10. Grigg, E.R.N., "The New History of Radiology," *Radiologic Technology*, 36(4):1965.
11. Hodges, Fred J., and Willis S. Peck, *Introduction to Radiology*, Edwards Brothers, Inc., Ann Arbor, Michigan, 1939.
12. Hoing, Margaret, "History of the ASXT: 1920-1940," X-Ray Technician, 17(6):1946.
13. Holland, Marilyn, "American Society of Radiologic Technologists" in Gurley and Callaway's (eds), *Introduction to Radiologic Technology*, Multi-Media Publishing, Inc., Denver, Colorado, 1982.
14. Jerman, Ed C., *Modern X-Ray Technic*, Bruce Publishing Company, St. Paul, Minneapolis, 1928.
15. "John Sanford Shearer: 1865 - 1922," *Medical Radiography and Photography*, 27(3):1951.
16. Hart, Alan L., *These Mysterious Rays*, 2nd ed, Harper and Brothers Publishers, New York, 1943.
17. Kassabian, Mihran K., *Roentgen Rays and Electro-Therapeutics*, J. P. Lippincott Company, Philadelphia, PA 1910.
18. Kraft, Ernest, "Wilhelm Conrad Roentgen: 1845 - 1923," *Medical Radiography and Photography*, 49(2):1973.
19. Lauer, O. Gary, "Radiography in the United States Army During World War I," *Radiologic Technology*, 56(6):1985.
20. McKnight, Neta, "American Registry of Radiologic Technologists," in Gurley and Callaway's (eds), *Introduction to Radiologic Technology*, Multi-Media Publishing, Inc., Denver, Colorado 1982.
21. "Mihran Krikor Kassabian: 1870 - 1910," *Modern Radiography and Photography*, 25(1):1949.
22. Milligan, Virginia, A., "The Anatomy of Change," *Radiologic Technology*, 48(3):1976.
23. Morton, S.A., "Registry Panel," *X-Ray Technician*, 24(2):1952.
24. Sante, L. R., and Harry W. Fischer, *Manual of Roentgenological Technique*, 12th ed., Edwards Brothers, Inc., Ann Arbor, Michigan, 1962.
25. Thompson, S. P., President's Address to British Roentgen Society, *Arch. Clin. Skiagraphy*, Vol. 2, Nov. 1897. Cited in Glasser, O. (Editor): *The Science of Radiology*, Springfield, Ill., Charles C. Thomas, 1933, p. 2. Cited in *The Fundamentals of Radiography*, 11th ed., Eastman Kodak Company, Rochester, New York, 1968.
26. *United States Army X-Ray Manual*, Paul B. Hoeber, New York, 1918.
27. *X-Ray Techniquiz*, General Electric Company, Milwaukee, Wisconsin, 1965.

Plate I

Principal Properties Of Radiographic Quality

PHOTO-RADIOGRAPHIC PROPERTIES

Radiographic Density

- Milliamperage
- Time
- Kilovoltage
- Distance

Radiographic Contrast

Subject Contrast

Exposure Latitude

- Kilovoltage
- Filtration
- Beam Restrictors
- Tissue Thickness/Composition

Film Contrast

Film Latitude

- Film Type
- Processing
- Screens vs. Non-Screen

GEOMETRIC PROPERTIES

Recorded Detail

Geometric Unsharpness

- Focal Spot Size
- Focal Film Distance
- Object Film Distance

Screen Unsharpness

- Film - Screen Contact
- Phosphor Layer Thickness

Motion Unsharpness

- Tube, Object, Film Movement

Distortion

Size (Magnification)

- Object Film Distance
- Focal Film Distance

Shape (True Distortion)

- Tube
- Object
- Film Alignment

PERCEPTUAL PROPERTIES

Radiographic Noise

Artifacts

- Sensitized
- Non-Sensitized
- Physical

Mottle

- Quantum
- Structural
- Film Graininess

Viewing Conditions

- Sight
- Environment
- Distance
- Display

Evaluation Criteria

- Photo-Radiographic
- Geometric
- Perceptual

Chapter 2

The Photo-Radiographic Properties of Image Visibility

Chapter Outline

1. Learning Objectives
2. Introduction
3. Radiographic Density
4. Radiographic Contrast
   A. Subject Contrast
      1. Exposure Latitude
   B. Film Contrast
      1. Film Latitude
5. Contrast and Latitude Terminology
6. Summary
7. Study Questions
8. Bibliography

Learning Objectives

Upon completion of this chapter, the student should be able to:

1. List the two conditions that must be present in a radiographic image that would indicate acceptable visibility of image details.
2. Define radiographic density and radiographic contrast.
3. Define subject contrast and exposure latitude. Explain their relationship to exposure error.
4. Define film contrast and film latitude. Explain their relationship to average gradient.
5. List three subjective and three objective adjectives that are used to describe contrast and latitude.
6. Recognize and identify short-scale contrast and long-scale contrast images and film characteristic curves.

Introduction

When a physician experiences difficulty viewing all the image details that should be present in a radiograph, the ability to conduct an optimum (best) diagnostic interpretation of the patient's condition may be significantly impaired. The radiographer, therefore, must realize that there is a difference between seeing an anatomical image on a radiograph and seeing one with the complete dimensions of the structural details within that image. For example, the skull in Figure 2-1A primarily visualizes the gross skull silhouetted against a black background. Although an image of a skull may be identified, the structural details of the internal image are NOT visible. Notice that visibility of the structural details of the orbits or maxillary sinuses is difficult. If the patient had trauma or pathology in these areas, the physician might not be able to accurately diagnose the condition.

A radiograph demonstrating acceptable visibility of image details will show the gross structures as well as the internal structural details of the anatomic image. Notice in Figure 2-1B that the radiographic density differences (contrast) of the orbits and maxillary sinuses are sufficient for the eye to easily distinguish one from another. Producing these distinguishing radiographic densities requires the radiographer to establish an appropriate exposure technique for a given radiographic imaging system (x-ray machine-tube, grid, screen-film combination, and film-processing system), based on the thickness and condition (trauma, pathology) of the anatomic part being radiographed. The imag-

Figure 2-1. Examples of (A) radiograph of the skull without visibility of skull details and (B) radiograph with optimum visibility of skull structural details.

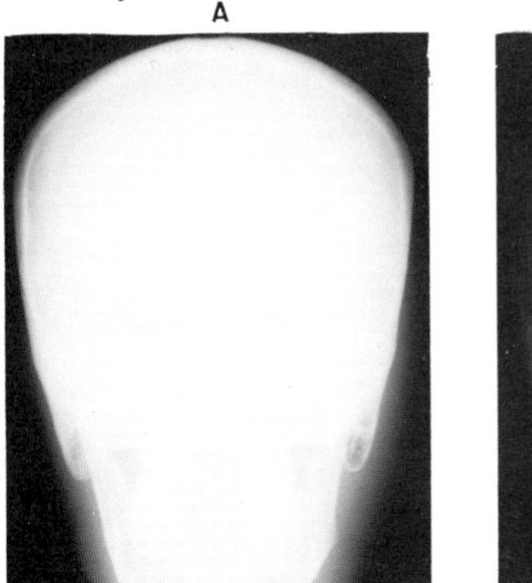

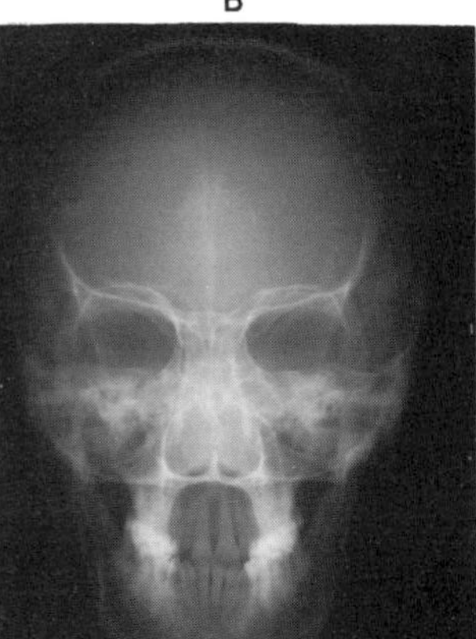

ing characteristics produced, as a result, will identify the quality of the photo-radiographic properties of image visibility; radiographic density (deposited black metallic silver), and radiographic contrast (distribution of black metallic silver). Radiograpic contrast may be divided into subject contrast—exposure latitude and film contrast—film latitude.

The authors prefer to describe the properties of image visibility as "photo-radiographic" in nature as opposed to "photographic." "Photo-radiographic" is a term that correctly associates an imaging process based on the *transmitted* effects of x-rays and the *reflected* effects of light from intensifying screens on x-ray film. Today, the overwhelming majority of radiographs are produced using intensifying screens in conjunction with x-ray film.

Radiographic Density

Radiographic density is a photo-radiographic property affecting image visibility and is defined as the quantity of blackness appearing on a radiograph. Obtaining the correct density of the structural details being radiographed is critical if quality images are to be achieved. Without radiographic density, radiographic contrast could not be realized. Without radiographic contrast, the image would not be visible. Therefore, radiographic density can be considered the principle imaging characteristic from which an understanding of radiographic quality may begin.

Radiographic Image Formation

The grey or black areas on a radiograph are, in actuality, aggregates of tiny black metallic silver crystals that exist within the emulsion of the imaging recording medium known as x-ray film (Figure 2-2). Briefly, x-ray film emulsion is composed of sensitive silver bromide crystals suspended in a gelatin substance that is spread evenly upon a transparent, blue-tinted, polyester support base (Figure 2-3). When energy strikes x-ray film, either from light emitted from intensifying screens, radiant, or chemical energy, electrons are released and migrate to a sensitivity speck located within the silver bromide crystal. At this sensitivity speck, these electrons react with other silver ions and form black metallic silver atoms. These deposited black metallic silver atoms allow developer solution to convert (reduce) the entire silver bromide crystal to black metallic silver. Thus a portion of the visible image is formed. This process is commonly known as the Gurney-Mott Theory of Latent Image Formation (Figure 2-4).

Under specified developing conditions those silver bromide crystals that were not struck by radiation are not reduced to black metallic silver atoms. (Given excessive time, temperature, or specific activity conditions of development, however, even these unexposed silver bromide crystals can become reduced.) Instead, they are cleared (removed) from the polyester film base due to the action of the fixer, the second chemical solution used in the processing cycle. As a result, the opposite observable imaging characteristic results, an absence of radiographic density. This is interpreted as white or clear when observed on the viewbox.

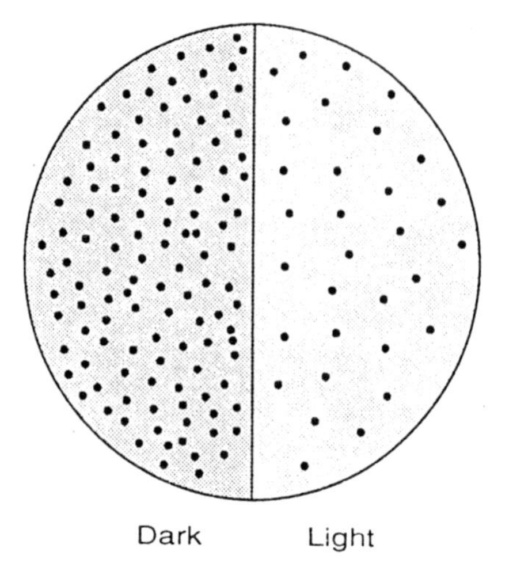

Figure 2-2. Diagrammatic representation of close-up view of radiographic density indicating compositon of aggregates of minute black metallic silver crystals.

Reciprocity of Densities

The quantity of silver bromide crystals reduced to black metallic silver atoms theoretically will be the same as long as the quantity of electromagnetic radiation reaching the film emulsion remains constant. This is the concept behind the *reciprocity law* which theoretically only applies to x-ray films that are exposed directly to x-rays as might be the case using cardboard or plastic film holders. When x-ray film is exposed to light from intensifying screens, the number of silver bromide crystals developed does not remain constant for extremely short or long exposures. This is the concept behind the *reciprocity law failure* and is primarily of concern for special procedure radiographers who use extremely short exposure times with intensifying screens. On the other hand, due to the improved characteristics of modern x-ray film emulsions, the failure of the reciprocity law is of no real concern to radiographers in general radiography. For all practical purposes, as long as the same type of exposure holder is used, equal exposure intensities (mAs) with differing mA and time combinations will approximate equal radiographic densities (Figure 2-5).

Densitometry

Measuring the amount of radiographic density on a radiograph is called densitometry and is done with a device known as a den-

Figure 2-3. Cross section of basic structure and composition of x-ray film.

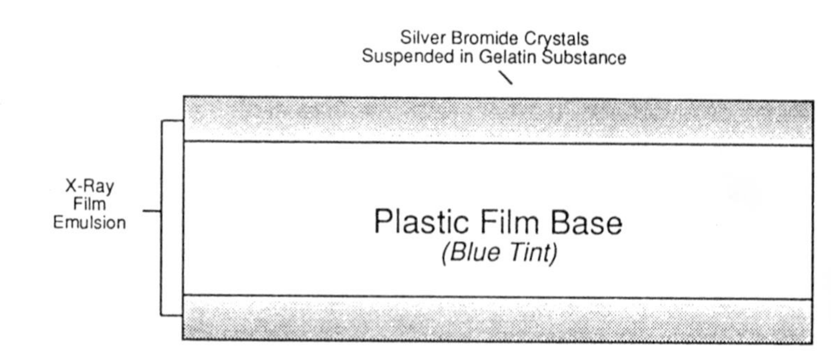

Figure 2-5. Example of application of the reciprocity law. Four different exposure settings (same mAs values) used with direct exposure film holders (non-screen) depict same radiographic densities.

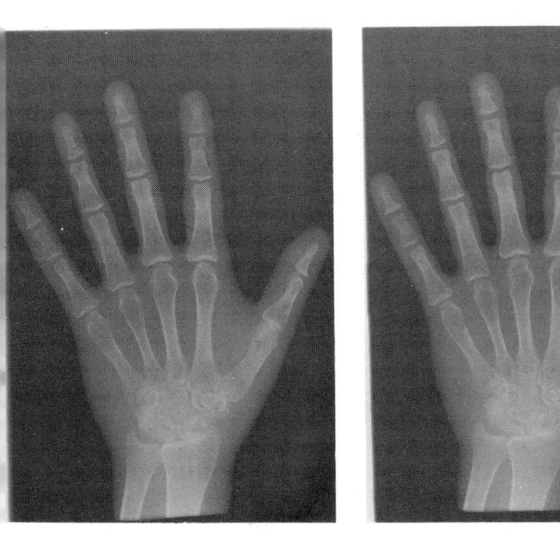

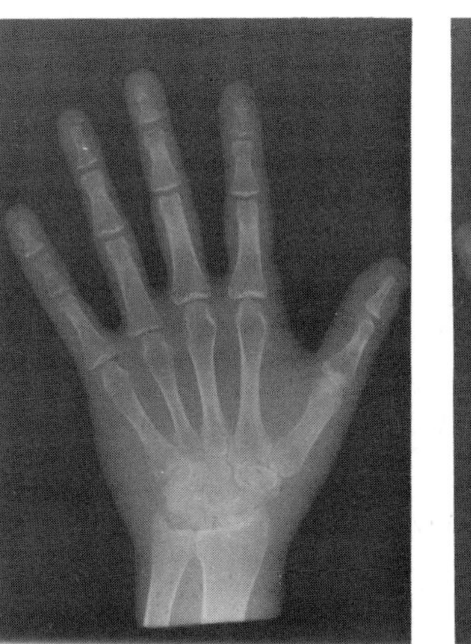

(400 mA at 1/4 sec) (200 mA at 1/2 sec) (100 mA at 1 sec) (50 mA at 2 sec)

sitometer (Figure 2-6). A densitometer is an electronic device containing a light source, a probe with a sensor for measuring light transmission through the radiograph and a logarithmic amplifier that converts the voltage generated by the light sensor into measurable units of optical density.

$$\text{Optical Density} = \text{Log}\,\frac{\text{Incident Light Intensity}}{\text{Transmitted Light Intensity}}$$

The measurement of the relationship between the intensity of exposure and the resultant density on the radiographic film is called sensitometry. When optical density measurements are plotted in relationship to relative exposure, a characteristic curve demonstrating the response of x-ray film emulsion to exposure or processing may be identified, (Figure 2-7). This curve is also known as the H & D curve after Hurter and Driffield who first described it as the ''sensitometric'' curve which is indicative of the sensitive properties of x-ray film to exposure and processing.

The range of optical densities that may be measured logarithmically by many commercial densitometers used in medical radiography is between 0.0 and 3.0 density units. Where one hundred percent (100%) of the light is transmitted through a point on the radiograph, an optical density measurement of 0.0 would be indicated. On the other hand, where one hundred percent (100%) of the light is prevented from passing through a point on the radiograph (because of a tremendous deposition of black

Figure 2-4. Conceptualization of Gurney-Mott Theory of Latent Image Foundation.

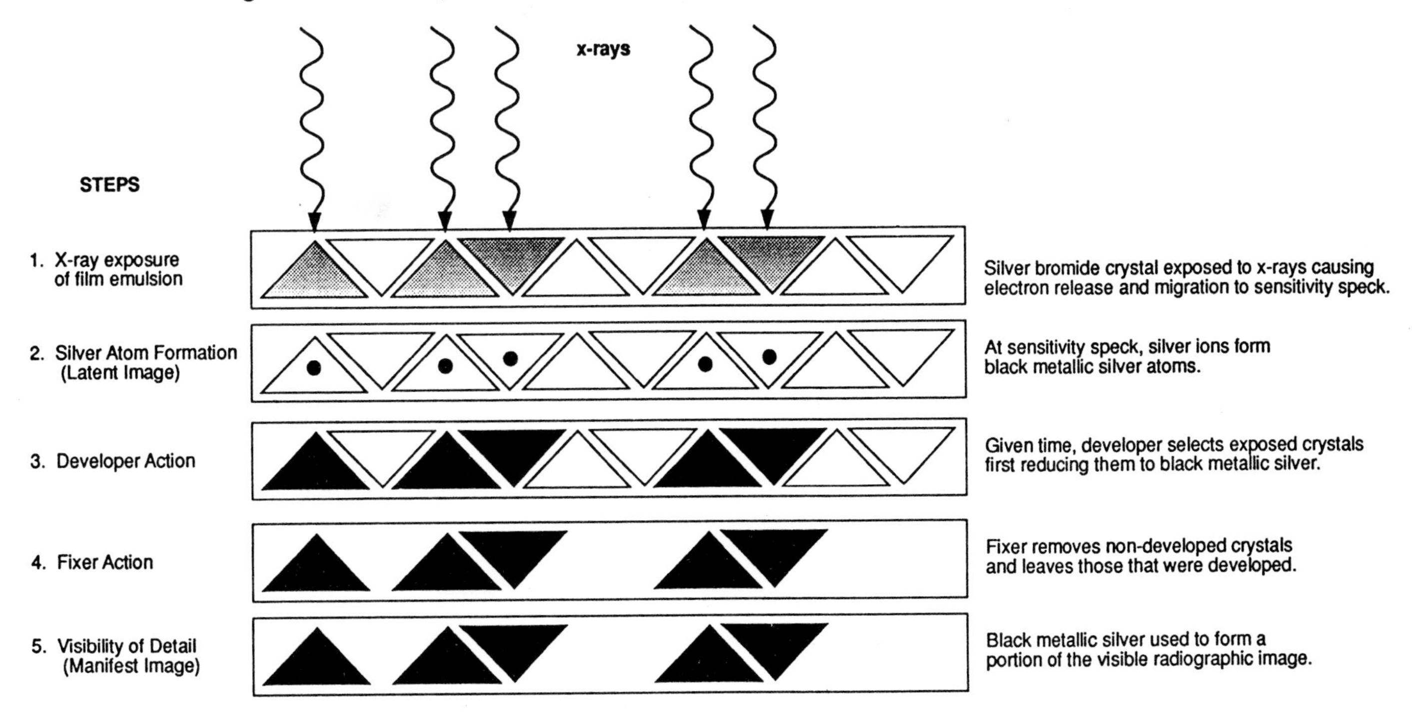

metallic silver crystals), an optical density measurement of 3.0 would be indicated on the densitometer. (Table 2-1)

Measurement of zero (0) density from a radiograph on the densitometer is not likely to happen because the polyester film base contains a blue tint and a tiny amount of fog that may be from exposures from darkroom safelights, processing, or scatter radiation. The blue tint is manufactured into the polyester film base to enhance contrast effects. Base plus fog (B + F) values are not the same for all radiographs, but acceptable levels should not vary much from .18 density units.

Diagnostic Viewing Range

The multitude of densities comprising the structural details of the radiographic image vary considerably between densitometric readings of B + F and 3.0. The range considered to be best (optimal) for visualization and therefore construction of the radiographic image is between 0.25 plus B + F and 2.0 plus B + F density units (Table 2-1). Generally, radiographic densities falling above or below this range provide such small measurable differences between adjacent densities that the ability of the eye to view image details is significantly impaired. Radiographs with excessive densities above this range or insufficient densities and below this diagnostic viewing range are exhibiting overexposed or underexposed conditions of the image details. This suggests the use of technical settings based on incorrect exposure guidelines, assuming all radiographic and processing equipment is operating efficiently.

In some cases, overexposed radiographs may be visualized by "hot lighting" them in order to afford the physician a better radiographic interpretation (Figure 2-8). This is elected instead of repeating another less-exposed radiograph since a repeated radiograph would increase the radiation dose to the patient as well as cause additional and unnecessary resource utilization. Nothing

Figure 2-6. Example of densitometer used in medical radiography.

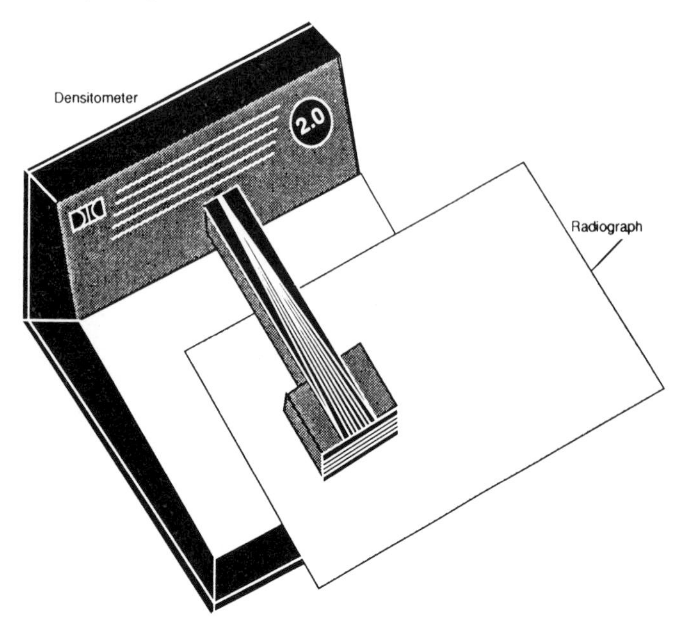

TABLE 2-1.
The Useful Diagnostic Density Viewing Range from Which the Visible Radiographic Image Is Constructed

Optical Density		Percent Light Transmission
0.00		100.0%
0.25		56.0%
0.50		31.0%
0.75		17.0%
1.00	Diagnostic Viewing Range (0.25–2.00)	10.0%
1.25		6.0%
1.50		3.0%
1.75		2.0%
2.00		1.0%
2.25		0.6%
2.50		0.3%
2.75		0.2%
3.00		0.0%

Figure 2-7. Anatomy of X-ray Film Characteristic Curve

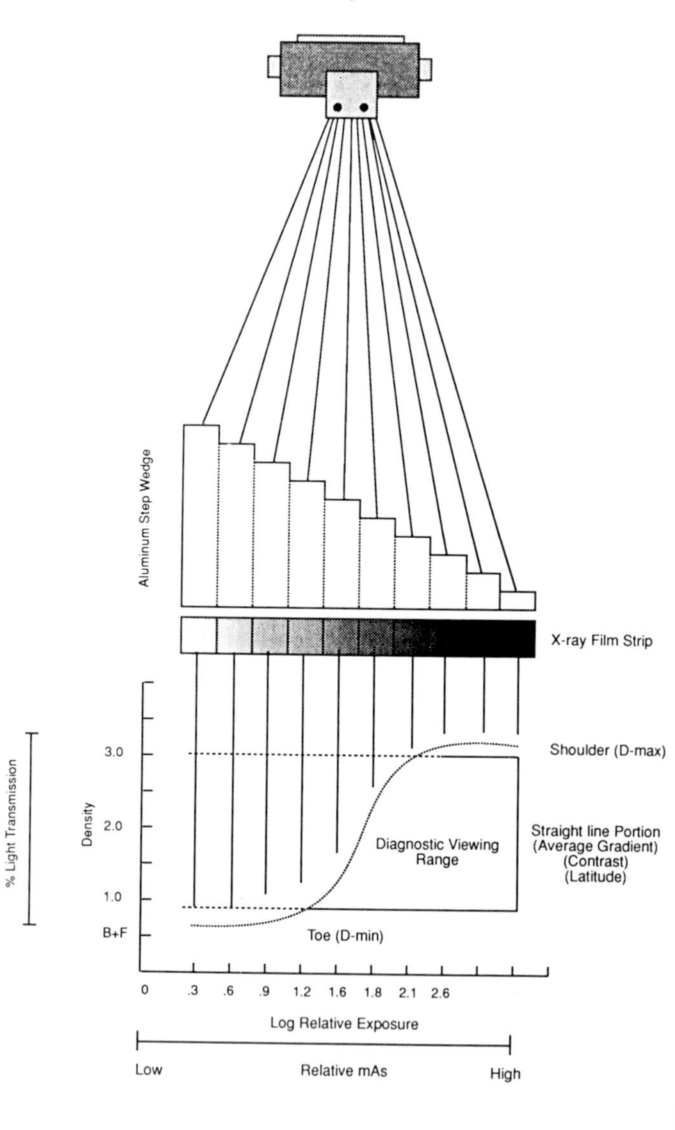

Figure 2-8. A "hot light" device used for visualizaton of overexposed radiographs.

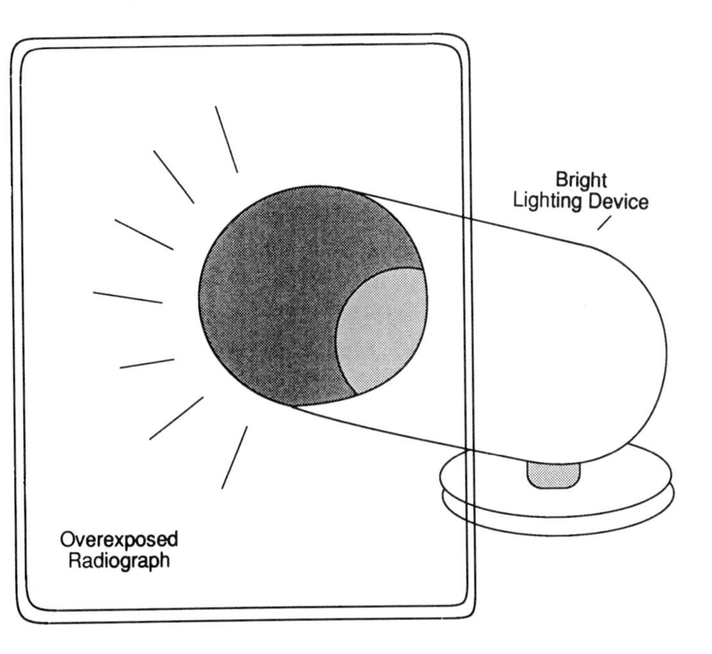

can be done, however, to salvage underexposed radiographs that are too light for a diagnostic interpretation except to repeat them.

Figure 2-9, illustrates three radiographs of a lateral skull. Radiograph A is underexposed and not acceptable for interpretation. Radiograph C is overexposed and not acceptable either. Radiographic densities comprising radiographs A and C represent the extremes of the diagnostic density viewing range. Radiograph B is correctly exposed since the details of the radiographic image are visible. It contains densities that fall primarily within the useful diagnostic viewing range.

Controlling Radiographic Density

Radiographers have learned to control the amount of radiographic density desired on a radiograph through the manipulation of four principal exposure variables that affect the "quantity" of x-rays produced. Usually the radiographer's first and second choice is between milliamperage or seconds, (mAs) since both directly control radiation output intensity. By doubling the mA or time setting at the control panel of the x-ray machine, one may double the amount of radiation and thus the amount of radiographic density desired on the film provided there is adequate penetration of the x-ray beam to pass through the part and reach the film. A third choice is kilovoltage (kV) since a 15 percent increase or decrease will have the same effect on x-ray film as if the mAs were doubled or halved. This is known as the 15 percent rule. A fourth choice is distance since the intensity of the x-ray beam increases inversely proportional to the square of the distance. This is the premise for the inverse square law. Other variables also have an impact on radiographic density such as part thickness and compositon, compression of part, respiration, secondary radiation, light, processing, grids, screens, and equipment efficiency. Many of these will be discussed individually in subsequent chapters.

Radiographic Contrast

Radiographic contrast is another photo-radiographic property affecting image visibility. It is broadly defined as the measurable or observed difference between adjacent radiographic densities. Obviously, without density differences, the structural details of the radiographic image could not be created for visualization. The totally black, grey, or white (clear) radiograph contains no image details and is thus void of any radiographic contrast. Variable tones from white to grey to black are required to portray anatomical images. When the measurable or observable differences between adjacent densities is great, the radiograph is considered to be one of high- or short-scale contrast. The general latitude of the radiograph is narrow with a short scale of intermediate tones. Low- or long-scale contrast on a radiograph contains a general latitude that exhibits a longer scale of tones from which the image details are constructed. Radiographic contrast is specifically divided into subject contrast (which is related to exposure latitude), and film contrast (which is related to film latitude).

Subject Contrast

Subject contrast is the result of the attenuation (absorption) of the x-ray beam by differences in tissue composition. This process is called differential absorption (Figure 2-10). X-rays are attenuated differently depending on the compactness or atomic number of the chemical elements that compose the various anatomical tissues and structures of the human body. Bone, for

Figure 2-9. Examples of (A) underexposed (B) diagnostically acceptable, and (C) overexposed skull radiographs.

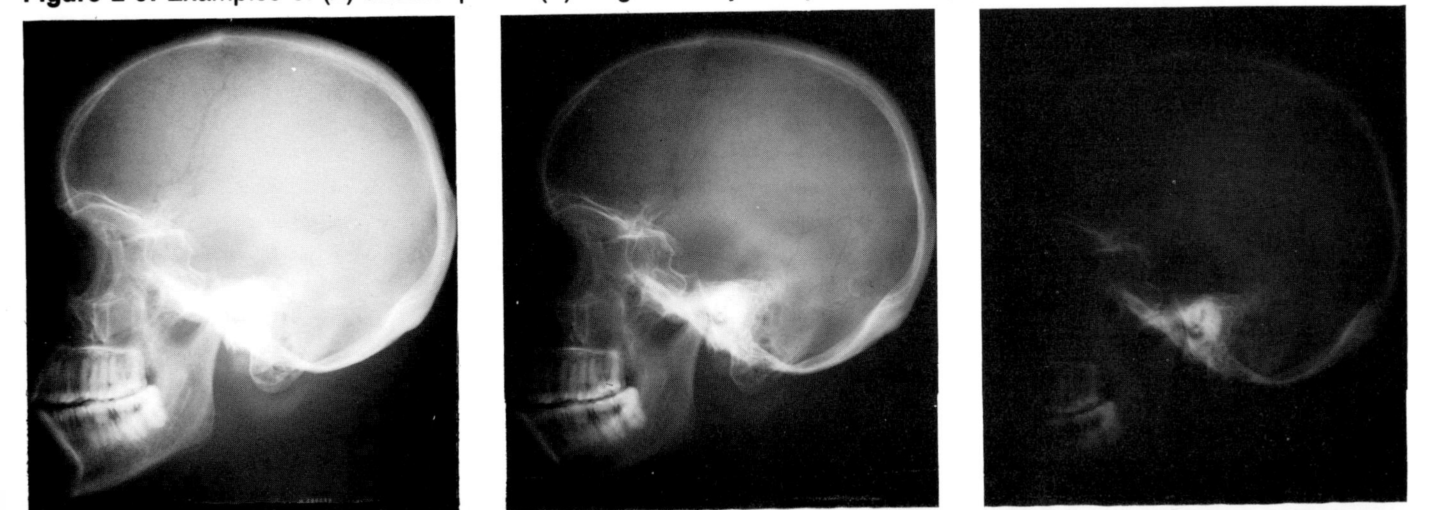

Figure 2-10. Conceptualization of differential absorption of x-rays by anatomy composition to form subject contrast.

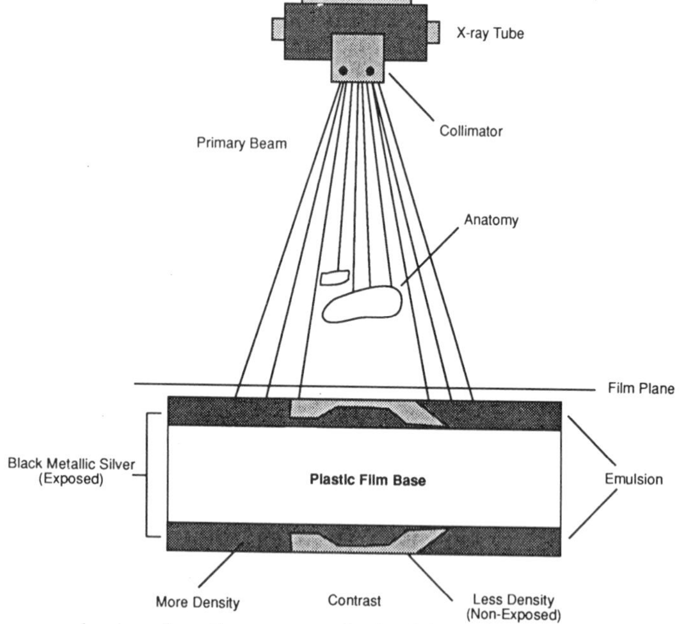

example, is primarily composed of calcium that has an atomic number of 20. This is considerably more dense than air that is composed of nitrogen and oxygen with respective atomic numbers of 7 and 8. Because bone is more dense than air, it will therefore attenuate more x-rays than a sinus cavity that is composed of air. The attenuating characteristics of bone prevent the x-rays from reaching the x-ray film and exposing the sensitive silver bromide crystals in the emulsion.

Anatomical tissues and structures that allow x-rays to easily penetrate are classified as radiolucent in nature. Those that easily attenuate are classified as radiopaque. The chest radiograph in Figure 2-11 illustrates both radiolucent and radiopaque tissue types. Notice that the greater radiographic densities correspond to the radiolucent lung field that is composed primarily of air filled aveólar sacs. On the other hand, the reduction in radiographic densities corresponds to the radiopaque bones of the ribs and thorax. Table 2-2 provides a classification of tissue types according to their radiolucency, radiopacity, and atomic number.

Figure 2-11. Chest radiograph demonstrating (A) radiolucent (lung-air) tissue, and (B) radiopaque (bone-muscle) tissues.

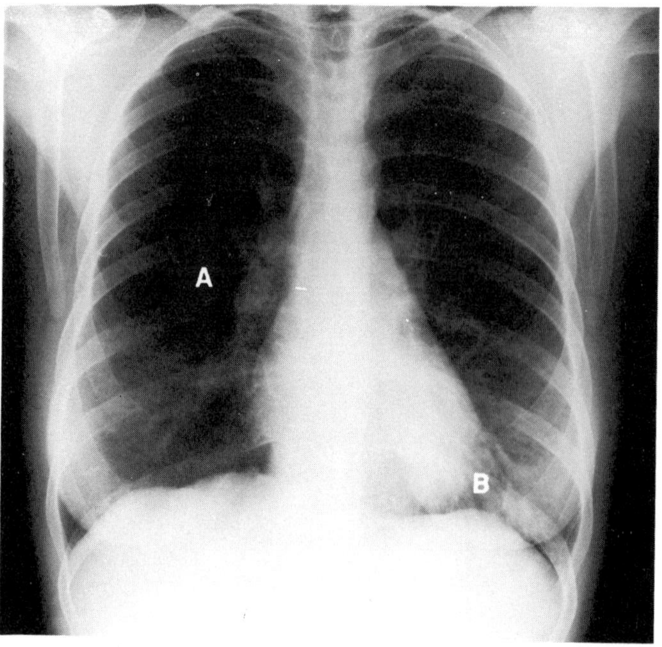

TABLE 2-2.
Classification of Tissue Types According to Their Radiolucency, Radiopacity, and Atomic Number.

Tissue Type	Radiolucency	Atomic Number
Lung (Gas)	Very Radiolucent	Low
Fat		
Muscle		
Blood		
Connective Tissue		
Cartilage		
Epithelium		
Gallstones		
Kidney Stones		
Bone		
Calcium Salts		
Metal	Very Radiopaque	High

Radiographers have learned to control subject contrast through the manipulation of principal variables that affect the "quality" of the x-rays produced for a given exposure. The exposure variable primarily used for this is kilovoltage (kV). When the x-ray penetration (kV) is high or when the tissue composition is homogeneous (uniform tissue density), the ratio of x-ray intensities emerging from the regions of the tissues is low (for example, flat grey tones). Such a condition defines a radiograph with long-scale subject contrast (Figure 2-12A). When the x-ray pentration (kV) is low or when the tissue composition is heterogeneous (nonuniform tissue density), the ratio of x-ray intensities emerging from the regions of the tissues is high (for ex-

Figure 2-12. Examples of long-scale (A) and short-scale (B) subject contrast. (The high kV technique used in radiograph A is designed to enhance heart and lung detail whereas the lower kV technique used to produce radiograph B is designed to enhance rib detail.)

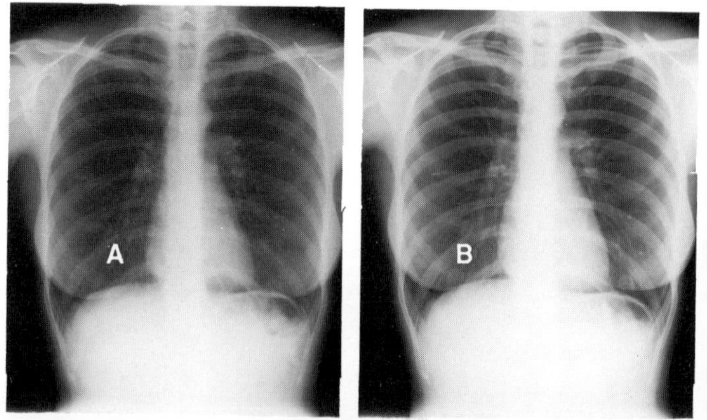

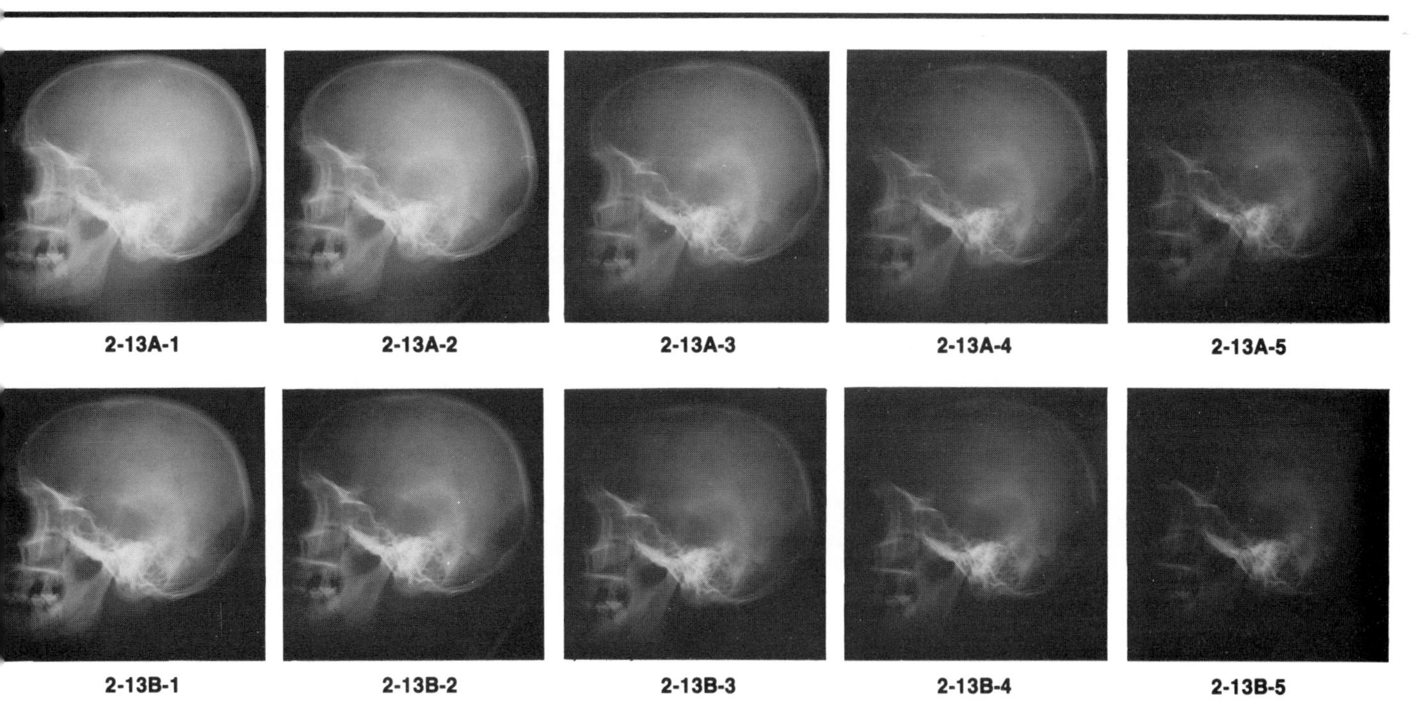

Figure 2-13. Exposure latitude of skull radiographs. Series of skull radiographs in top row represents wide exposure latitude and less chance of exposure error. Series of radiographs in bottom row represents narrow exposure latitude and greater chance of exposure error.

ample, black-and-white tones). Such a condition defines a radiograph with short-scale subject contrast (Figure 2-12B).

Exposure Latitude

Exposure latitude is defined as the range of exposures (mAs), that may be used to obtain a radiograph that demonstrates acceptable visibility of image detail for a particular anatomical part based on a given pentration (kV) quality. Exposure latitude is associated with subject contrast in the following manner. When penetration quality (kV) of the x-ray beam is high, a low-contrast image is realized and thus exposure latitude widens (Figure 2-13A). That is, the range of acceptable exposures that may be used to affect density (mAs) for the production of a series of acceptable radiographic images of a part increases. When the penetration of the x-ray beam is at lower values for a given anatomical part, a high-contrast image is realized and exposure latitude narrows (Figure 2-13B). That is, the range of acceptable exposures that may be used to affect density (mAs) for the production of a series of acceptable radiographic images decreases.

''Exposure error'' is the term used to describe the probability that a radiographic exposure will result in unacceptable visibility of image details. Under highest acceptable penetration (kV) conditions, exposure latitude widens and the number of acceptable radiographs produced for a given exposure (mAs) increases. The probability of an exposure error decreases since there is a greater range of exposures that may be selected for the production of an acceptable radiographic image. On the other hand, if less than acceptable penetration values are used, exposure latitude narrows and the number of exposures (mAs) resulting in the creation of acceptable radiographic image details decreases. Thus the probability that exposure error will be experienced increases (Table 2-3).

TABLE 2-3.
The Relationship Between X-Ray Penetration (kV), Subject Contrast, Exposure Latitude, and Exposure Error.

Penetration (kV)	Subject Contrast	Exposure Latitude	Exposure Error
Highest Acceptable	Low (Long Scale)	Wide (Increases)	Decreases
Lowest Acceptable	High (Short Scale)	Narrow (Decreases)	Increases

Radiographers that become aware of the relationships between exposure latitude and exposure error are able to vary the kind of subject contrast desired with the application of the ''15 Percent Rule.'' For example, let's say that a radiographer wants to produce a radiographic image of some anatomic part that demonstrates long-scale subject contrast while maintaining the same radiographic density based on an original radiographic exposure technique of 70 kVp at 50 mAs. By adjusting penetration of the x-ray beam to higher acceptable limitations, the subject contrast will become lower (longer scale) according to Table 2-3. A 15 percent increase in the original kV will have an effect on radiographic density as if it were doubled. Therefore, 15 percent of 70 kV is an additional 10.5 kV and this results in a final kV increase of approximately 81 kV (70 kV + 11 kV = 81 kV). Next, in order that the original radiographic density is essential-

Figure 2-14. Calculation of average gradient from the slope of the straight line portion of the characteristic curve to demonstrate measurement and comparison of film contrast and film latitude. (A, high contrast/narrow latitude; B, intermediate; C, low contrast/wide latitude.)

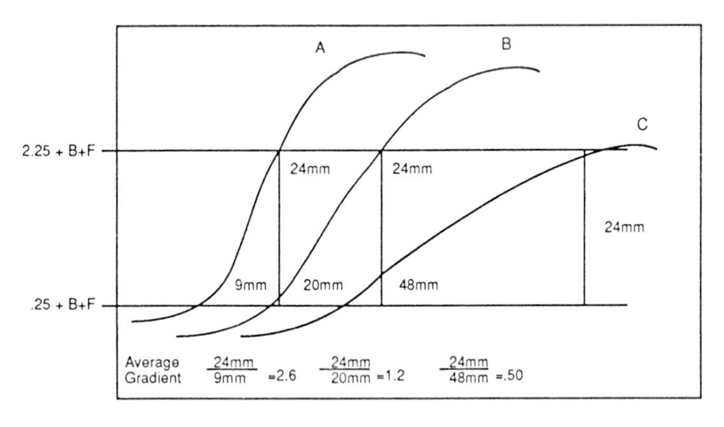

ly maintained, a corresponding halving of the mAs needs to take place. Therefore, the radiographic density of 50 mAs at 70 kV would be equivalent to the radiographic density of 25 mAs at 81 kV; the new radiographic exposure technique resulting in a change of subject contrast from short scale to long scale.

Film Contrast

Film contrast pertains to the difference between adjacent radiographic densities based on the response of a given x-ray film "emulsion" type to exposure and processing. It may be associated with the average gradient as calculated from the straight line portion of the characteristic curve (Figure 2-14). Film contrast varies depending on the film manufacturer's preference for engineering long- or short-scale contrast characteristics within the x-ray film emulsion for a given radiographic processing chemistry. Consequently, a variety of x-ray films and processing chemistries are commercially available depending on the specific radiographic requirements.

The graphic representation of this relationship for three hypothetical x-ray films exposed to identical x-ray intensities is illustrated in Figure 2-14. Notice that the diagnostic viewing range (straight-line portion) for film A is more verticle with a slope (average gradient) of 2.6. The more vertical the straight-line portion of the characteristic curve, the greater will be the average gradient number. This is indicative of an x-ray film emulsion type (or developing chemistry) that is engineered to produce radiographic images exhibiting shorter scale contrast images based on a given x-ray exposure and processing chemistry. Curve C on the other hand represents an x-ray film that is manufactured for longer scale contrast characteristics for a given exposure and processing chemistry (average gradient = .50). Curve B represents a film type that exhibits film contrast characteristics between film types A and C (average gradient = 1.2).

There are several variables that will affect film contrast as depicted by measurement of the average gradient from th straight-line portion of the characteristic curve. For example, us of intensifying screens will generate higher film contrast com pared to direct exposure film holders (cardboard or plastic cassett holders). Similarly, increases in the time, temperature, or specifi activity of development will result in the film exhibiting its highe film contrast characteristics. With respect to film contrast, th responsibility of the radiographer is to exercise maximum con trol in establishing proper x-ray exposure techniques and pro cessing parameters that are designed to take full advantage o the film contrast specifications set by the film manufacturer. Thi may be facilitated when x-ray machines, ancilliary equipment film-processing equipment and chemistry are operating with high degree of efficiency in relationship to valid and reliable ex posure guides for a given anatomical part thickness o composition.

Film Latitude

Film latitude is an expression of film contrast. It is anothe quality that is engineered by the manufacturer into the film emul sion. Film latitude is defined as the ability of the x-ray film t respond to a range of exposures and record them as useful den sities (Jenkins: 1980). If short-scale film contrast refers to a film type that exhibits greater measurable differences between adja cent recorded densities, then the film latitude would be narrow That is, fewer radiographic density tones would be exhibited an the range of useful densities would decrease. On the other hand a film type that exhibits long-scale film contrast will have fewe measurable differences between adjacent densities and thus, th film latitude will be wide. That is, more radiographic density tone would be exhibited and the range of useful densities would in crease (Table 2-4).

TABLE 2-4.
The Relationship Between Film Contrast, Film Latitude and the Diagnostic Viewing Range

Film Contrast	Film Latitude	Diagnostic Viewing Rang
Short Scale (High)	Narrow	Decreases
Long Scale (Low)	Wide	Increases

Contrast and Latitude Terminology

In a typical radiography department or classroom, one may hea a variety of terms being used to describe various aspects of con trast and latitude. A radiologist may indicate that a radiograp needs to be repeated with *more* contrast. Radiographers evaluatin their radiographs for radiographic quality may indicate that th latitude of a particular film is *greater* than that of another. I the classroom or laboratory, educators and students may b discussing how to increase contrast while at the same time *reduc* exposure latitude. New radiography students can get very cor fused when a practical understanding of the concepts for cor trast and latitude are not clearly defined and understood. Tabl

2-5 is provided to assist in alleviating some of the confusion concerning the terminology of contrast and latitude.

TABLE 2-5.
Examples of Adjectives that Are Frequently Used to Objectively Describe Changes in Contrast and Latitude

Subject or Film Contrast		Exposure or Film Latitude	
Short Scale	Long Scale	Narrow	Wide
High	Low	Decreased	Increased
Increased	Decreased	Less	More
Greater	Lesser	Reduced	Expanded
More	Reduced		

In describing the contrast and latitude characteristics of an image or film type, one may objectively identify the specific direction these characteristics have taken using adjectives similar to those identified in Table 2-5. However, just about the time that one begins to understand these adjectives, someone begins to discuss contrast and latitude using subjective terms such as good, bad, improved, poor, fair, better, awful, nice or enhanced. Such terms may be extremely confusing, especially to the impressionable student radiographer, because they do not indicate the direction of change in contrast or latitude. Use of such subjective terms may improperly reinforce the concepts associated with the principles of radiographic quality, especially when they are used without making adequate comparisons between two or more radiographic images or x-ray film characteristic curves.

SUMMARY

The photo-radiographic property of radiographic quality pertains to the visibility of image details. When density differences between anatomic structures are sufficient for the eye to easily distinguish one from another, optimum visibility of image details may exist. The photo-radiographic properties of image visibility are radiographic density and radiographic contrast (Table 2-6). Radiographic density pertains to the amount of film blackening while contrast pertains to the measurable or observable differences between two or more adjacent densities. Contrast is further divided into subject and film contrast. Subject contrast pertains to density differences resulting from differential tissue absorption. Film contrast pertains to density differences engineered into the film emulsion based on specifications of the film manufacturer. Exposure latitude is associated with subject contrast and defines the range of exposure intensities (mAs) for a given beam penetration (kV) that may be used to obtain a radiograph with optimum visibility of image details. Less than optimum x-ray beam penetration results in shorter scale subject contrast, narrow exposure latitude, and greater exposure error. Film latitude is associated with film contrast and is the ability of the film to respond to a range of exposures and record them as useful densities. Given two different film types, identically exposed and processed, the one with a greater average gradient slope will exhibit shorter scale film contrast and narrow film latitude characteristics. The terminology describing contrast and latitude should be objective in nature indicating the direction of change; for example short scale, long scale, narrow, wide, increase, decrease, more, or less. Terms like good, bad, poor, or fair, are subjective and tend to confuse rather than explain directional changes in contrast and latitude.

TABLE 2-6.
Summary of the Principal Factors Affecting the Photo-Radiographic Properties of Image Visibility.

RADIOGRAPHIC DENSITY	RADIOGRAPHIC CONTRAST	
	Subject Contrast (Exposure Latitude)	Film Contrast (Film Latitude)
Miliamperage (mA)	Kilovoltage	X-Ray Film
Exposure Time (s)	Filtration	Processing
Kilovoltage (kV)	Beam Restrictors	Screens vs Non Screen Exposure
Distance	Grids	
	Tissue Thickness & Composition	

STUDY QUESTIONS

1. Define radiographic density, subject contrast, film contrast, exposure latitude, and film latitude.
2. Explain the Gurney Mott Theory of Latent Image Formation.
3. Explain the reciprocity law and reciprocity law failure.
4. What is a densitometer?
5. What are the values of the useful diagnostic density viewing range as recorded by a densitometer?
6. What is Base + Fog? Give an approximate acceptable B + F value as measured by a densitometer.
7. The radiographic density plus Base + Fog on an area of a radiograph measures 2.41. Calculate the Base + Fog value if the radiographic density is 2.25.
8. Which of the following x-ray film characteristic curves in Figure 2-14 is an indication of short-scale film contrast? Narrow-film latitude? Low-film contrast? Increased-film latitude?
9. The average gradient of a characteristic curve is calculated from the useful diagnostic density range of the straight-line portion of the characteristic curve. True? False?
10. A radiographic image that exhibits *excellent* contrast is one that also exhibits long-scale contrast. True? False?

BIBLIOGRAPHY

1. Bushong, Steward C., *Radiologic Science for Technologists: Physics, Biology, and Protection*, 2nd Ed.: C. V. Mosby Co., St. Louis, 1980.
2. Cahoon, John B.; *Formulating X-Ray Techniques*, 6th Ed.: Duke University Press, Durham, North Carolina, 1974.
3. *DensitometerModel TBX: Operating and Service Manual/Parts List*, Tobias Associates, Inc., Ivyland, Pennsylvania, 1975.
4. Goodwin, Paul N., Edith H. Quimby, and Russell H. Morgan, *Physical Foundations of Radiology*, 4th Ed.: Harper & Row, New York, 1970.
5. Jenkins, David, *Radiographic Photography and Imaging Processes*, University Park Press, Baltimore, Maryland, 1980.
6. Kirby, Cynthia, C., W. E. J. McKinney, and T. T. Thompson, *A Guide for*

Automatic Radiographic Processing and Film Quality Control, The American Society of Radiologic Technologists, Chicago, 1975.
7. Lamel, David A., et al, *The Correlated Lecture Laboratory Series in Diagnostic Radiological Physics*, U.S. Department of Health and Human Services, Rockville, Maryland, 1981.
8. Meschan, Isadore and D. J. Ott, *Introduction to Diagnostic Imaging*, W. B. Saunders, Philadelphia, Pennsylvania, 1984.
9. *The Fundamentals of Radiography, 12th Ed.: Eastman Kodak Co., Rochester, New York, 1980.*

Chapter 3

The Geometric Properties of Image Formation

Chapter Outline

1. Learning Objectives
2. Introduction
3. Recorded Detail
   - A. Geometric
   - B. Screen
   - C. Motion
4. Distortion of Image Details
   - A. Size Distortion (Magnification)
     - 1. Calculations
       - a. Magnification Factor
       - b. Image Width
       - c. Percentage of Magnification
   - B. Shape Distortion
5. Summary
6. Study Questions
7. Bibliography

Learning Objectives

Upon completion of this chapter, the student should be able to:

1. Define recorded definition of detail.
2. Define penumbral and umbra effect.
3. Calculate the penumbral effect of geometric unsharpness.
4. List two factors responsible for screen unsharpness.
5. List three factors responsible for motion unsharpness.
6. Define size distortion and shape distortion.
7. Calculate magnification factor, image width, and percentage of magnification.

Introduction

In addition to creating a visible radiographic image of the anatomical structural details through the proper control over radiographic density and contrast, the radiographer must also be able to create one that is geometrically balanced. That is, the recorded image details must be sharply defined and formed in the proper anatomical representation. This chapter will focus on two properties of the geometric formation of the radiographic image. They are *recorded definition* and *distortion* of image details. The recorded definition of image details refers to

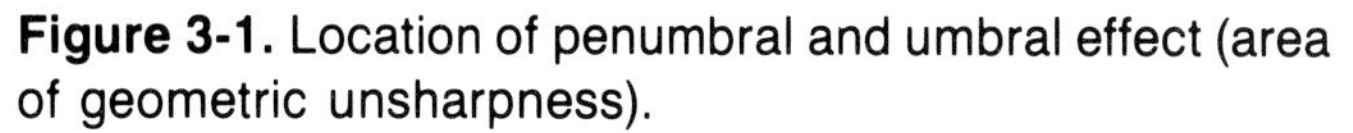

Figure 3-1. Location of penumbral and umbral effect (area of geometric unsharpness).

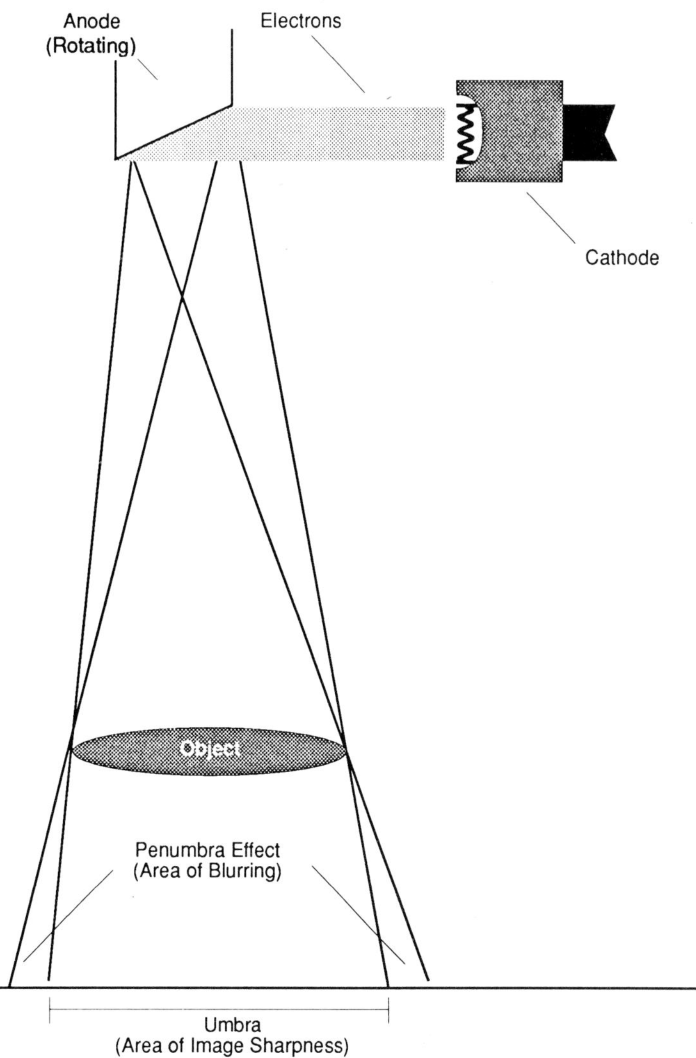

geometric, screen, and motion unsharpness. Distortion of image details refers to the size and shape of the anatomic image.

Recorded Detail

Closer scrutiny of the structural details of the radiographic image will reveal whether or not they are focused or sharply defined. When the parameters of the external and internal structural edges of image details do not appear out-of-focus with a fuzzy, hazy, or blurred appearance, the recorded details are considered to be sharply defined and acceptable enough so that an optimum diagnostic interpretation may be made by the physician. Out-of-focus images lack recorded definition of detail.

The human eye is capable of detecting unsharpness in a radiograph when the distinct lines of structural details are wider than 0.2 mm (5 lines per millimeter). The major contributing element responsible for destroying the ability of the x-ray film to optimally record sharply defined detail is the penumbral effect. The area of unsharpness of image details is known as the *penumbra* while the area of sharpness identifies the *umbra*. (Figure 3-1) Three imaging characteristics may be observed on a radiograph as a source of unsharpness (penumbra) of image details. These include geometric, screen, and motion unsharpness.

Geometric Unsharpness

Geometric unsharpness is defined as the unsharpness in a radiographic image due to the relative distance between the focal spot, object being radiographed, and film. Radiographers have learned to control for minimal penumbral effect of geometric unsharpness by using small focal spots, long focal-film distances, and minimal object-film distances. Geometric unsharpness (penumbral effect) may be calculated using the following formula:

Figure 3-2. Calculation of penumbral effect using geometric unsharpness formula.

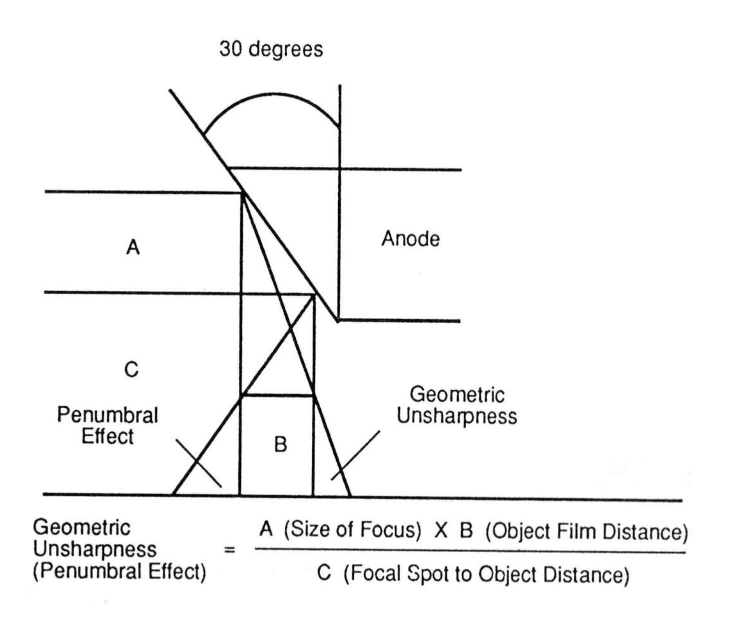

Screen Unsharpness

The unsharpness qualities present in a radiographic image as the result of the phosphor layer thickness of intensifying screens or from poor film-screen contact is known as screen unsharpness. The imaging recording materials used in conventional medical radiography are x-ray film and direct-exposure film holders such as cardboard or plastic cassettes, or indirect exposure film holders such as intensifying screen mounted in cassettes. For all practical purposes, the particular film type employed in conventional medical radiography does not pose any observable problems with respect to image unsharpness as long as one film type is used. However, the selection and use of the variety of intensifying screen speeds that may be available in radiology departments can present some image resolution problems (Table 3-1).

TABLE 3-1.
Image Resolution Properties of Four Conventional Intensifying Screens Compared to Screen Film in Direct Exposure Holders.

Type of Exposure Holder (Screen)	Resolution
Super Fast	0.55 mm of unsharpness
High Speed	0.45 mm of unsharpness
Par Speed	0.35 mm of unsharpness
Slow Speed	0.25 mm of unsharpness
*Screen Film in Direct-Exposure Holders	100 Line Pairs per mm of unsharpness

*Screen film exposed directly has approximately twice (55%) the resolving power compared to Super Fast Speed Screens.

Conventional (non rare earth type) intensifying screens made from, for example, barium sulfate, barium fluorochloride, or calcium tungstate phosphors are classified according to four relative speeds. Slow-speed screens are also known as detail or high-resolution speed screens. These screens are designed for optimum definition of image details. They are the next best thing to using screen-type x-ray film in a direct exposure holder. However, they require longer exposure times compared to other intensifying screens. As a result, they are not necessarily selected for use when patient motion is a problem. Par-speed screens are also known as medium or average speed intensifying screens. These screens are a general purpose screen providing adequate definition of image details using average exposure times. High-speed screens are also referred to as fast. This is the most widely used intensifying screen type in medical radiography at present due to its relatively low cost and adequate imaging capabilities. The fourth intensifying screen with the fastest possible exposure time capability is known as super fast or ultra high speed. Super-fast intensifying screens are the ultimate choice for stopping motion of the part. However, compared to all other conventional screens they result in the greatest loss of recorded detail.

Rare-earth type screens are made from elements such as adolinium, lanthanum, and yttrium. These types of intensifyng screens have the advantage of being the fastest available with he thinnest phosphor layer thickness resulting in minimal light iffusion and thus excellent definition of image details. Many epartments of radiology have converted from conventional calium tungstate screens to rare earth screens with compatible film. he result is good images with even less radiation than ultra high peed calcium tungstate screens.

Examples of representative intensification factors that indicate elative screen speed may be calculated when the exposure mAs sed without screens is divided by the exposure mAs used with particular screen type for a particular radiographic image. Table -2 provides a representative example of intensification factors alculated for the various screen speed types, including rare-earth ype screens. However, the reader is cautioned against assumng that these relative speed screen factors are exact or 100 perent reliable measurements since screens classified as fast, for xample, at a particular radiation exposure energy (kilovoltage) an respond differently when a different kilovoltage is used. At igher kilovoltages, screens give off more (brighter) light. In eneral, these intensification factors may be assumed as average or a broad range of exposure energies. Their use in calculating xposure conversions from one screen to another produces aceptable results, in most situations.

TABLE 3-2.
Relative Intensification Factors for Rare-Earth Type and Conventional Intensifying Screens Compared to Screen-Type Film in Direct-Exposure Holders.

Screen Type	Relative Intensification Factor
Rare-Earth Type	0.10
Super-Fast Speed	0.35
High Speed	0.50
Par/Medium Speed	1.00
Slow/Detail	2.00
Screen Film in Direct-Exposure Holders	20.00

For example, High-Speed Screens give off twice the light of Par-Speed Screens therefore the intensification factor for High-Speed Screens (relative speed) is half, .50, that of Par, 1.00.

The use of screen film in direct-exposure holders produces minimal image unsharpness because there is no light diffusion pattern from screen phosphors as is the case when using intensifying screens. This light diffusion pattern may be minimized depending on the type of intensifying screen employed such as slow, medium, high, super fast, and rare earth. Except for the relatively thin rare-earth type screens, usually the faster conventional screen speeds give off more light because of *thicker* phosphor layers (Figure 3-3). Thus, there is a greater light diffusion pattern generated. This light diffusion pattern may be ad-

igure 3-3. Conceptualization of light diffusion pattern based on screen speed as a result of phosphor layer thickness. Notice how faster screens, which have thicker phosphor layers, produce a greater light diffusion pattern.)

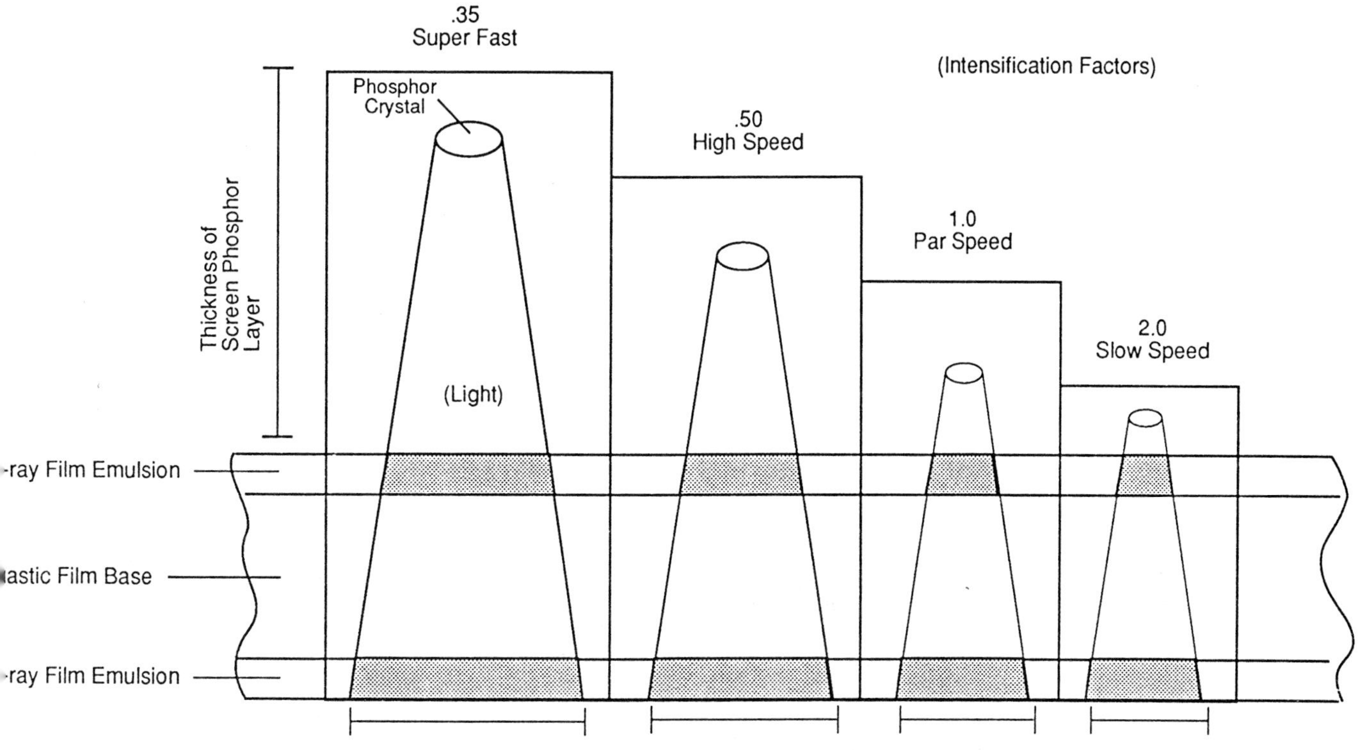

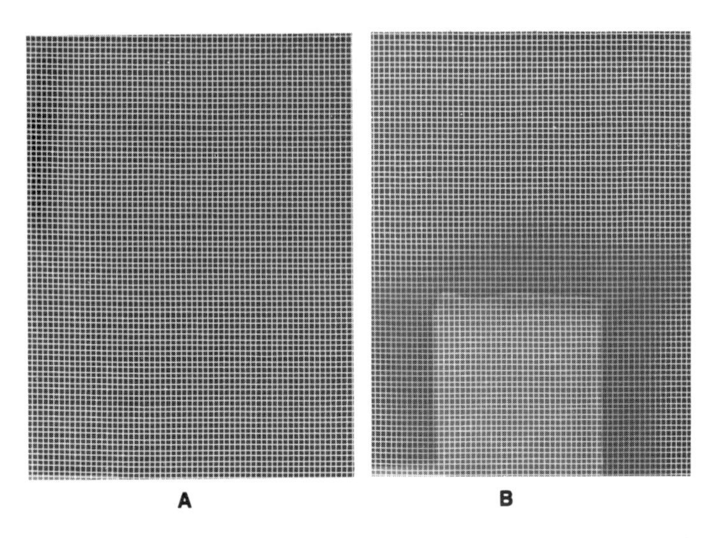

Figure 3-4. Comparison of image sharpness as a result of optimum (A) and poor (B) film screen contact.

ditionally exaggerated if there exists a condition known as "poor-screen contact" in which the intensifying screen(s) is prevented from coming directly into contact with the x-ray film emulsion surface (Figure 3-4). Close film-screen contact contributes to sharply defined radiographic images.

Motion Unsharpness

Motion unsharpness is defined as the loss of recorded detail in a radiographic image as a result of movement of the x-ray tube, patient, or recording medium. Radiographers have learned to control the effects of motion unsharpness by using reliable immobilization techniques (Figure 3-5) in conjunction with short-exposure times that are designed to minimize voluntary and involuntary patient motion during exposure. An example of voluntary patient motion in the production of a chest radiograph is the breathing of a patient during exposure. Even though the patient is asked to stop breathing for the duration of exposure, involuntary motion will continue within the patient in the form of the heartbeat. In order to further stop this involuntary heart motion, the shortest possible exposure time should be used. Since an adult heart beats an average of 60 times a minute, the safest exposure time recommended for completely stopping involuntarily cardiac motion should not exceed 0.04 (1/25) seconds (Donahue, 1980).

Figure 3-5. Examples of anatomical sponge block immobilizing devices.

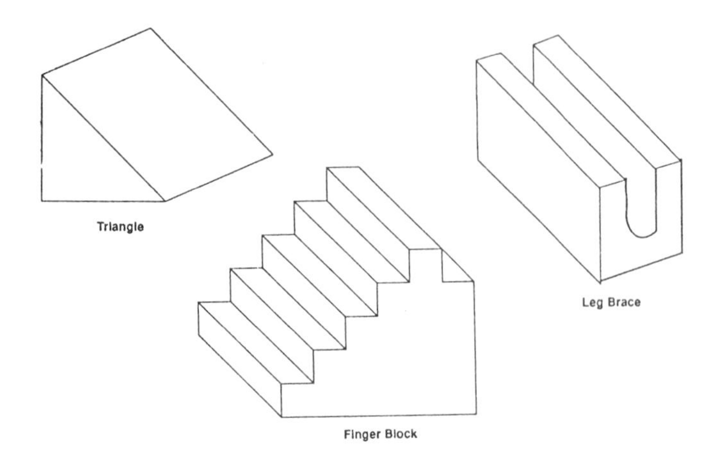

High-speed intensifying screens allow for the use of very short exposure times that may stop patient motion, but because the phosphor layers of high-speed screens may be thicker (except for rare-earth type screens) compared to screens with slower speeds, screen unsharpness will occur and result in additional screen unsharpness of image details. Use of rare-earth type screens, on the other hand, may allow for extremely shorter exposure capability that will further stop motion and provide fairly good definition. However, chances increase for the production of quantum mottle as a result of uneven distribution of x-rays (poor statistics) which will have a negative effect on recorded definition of image details. Radiographers can decrease the amount of motion resulting from tube or film movement by making sure that these devices are appropriately locked or secured with tape or supportive devices.

Distortion of Image Details

Radiographers strive to record image details in their truest size and shape; that is, without any annoying distortion qualities. Size distortion (magnification) and shape distortion (true distortion) are two aspects of the property of distortion that will now be examined.

Size Distortion

Size distortion is defined as the radiographic enlargement of the actual size of the structural details of the radiographic image due to increased object-film, decreased focus to object, or decreased focal film distance. In virtually all routine radiographic procedures, radiographers strive to reproduce image details that represent the actual size of the anatomical parts being radiographed. Under normal patient-care conditions, this is easily achieved by placing the part to be radiographed as close as possible to the film. Other times, especially during trauma or portable bedside radiography, patient-care circumstances may interfere with this process and enlargement of the structural details may be realized. Grossly enlarged details (without the use of fractional focal spots), are usually not tolerated since with enlargement of the part comes significant degrees of geometric unsharpness. On the other hand, the reader should be aware of a special magnification radiographic technique known as macroradiography. This is a technique in which the structural details of the radiographic image are deliberately magnified using fractional focusing x-ray tubes (0.3 mm or less) at increased object film distances (Selman, 1976). In this way, the radiographic image can be enlarged while retaining adequate definition of image details. Since smaller focal spot sizes result in the least geometric unsharpness, radiographers should realize today (using modern fractional focusing x-ray tubes) that focal film distances greater than the traditional 40 inches can be used to effectively decrease ometric unsharpness. Knowing this, one is able to maximize ficient use of the radiographic apparatus.

agnification Calculations

The *magnification factor* of a radiographic image may be lculated by constructing the ratio of the width of the diographic image to the actual width of the object being diographed (Figure 3-6). However, before making this calcula-on, one needs to first determine the *image width* that will be ojected on the x-ray film by the object being radiographed. This n be accomplished by applying the geometric relationship of milar triangles (Figure 3-7). (The same law of image forma-on is used to find the width of the object being radiographed ch as pelvic measurements during pelvimetry, when the im-ge width, focus-film distance, and object film distance are nown.) Once image width has been calculated, then the *percen-ge of magnification* of the enlarged radiographic image in rela-onship to its actual size may be calculated (Figure 3-8).

igure 3-6. Calculation of the magnification factor.

$$\text{Magnification Factor} = \frac{\textit{Image Width (in inches)}}{\textit{Object Width (in inches)}}$$

igure 3-7. Example of calculation of image width.

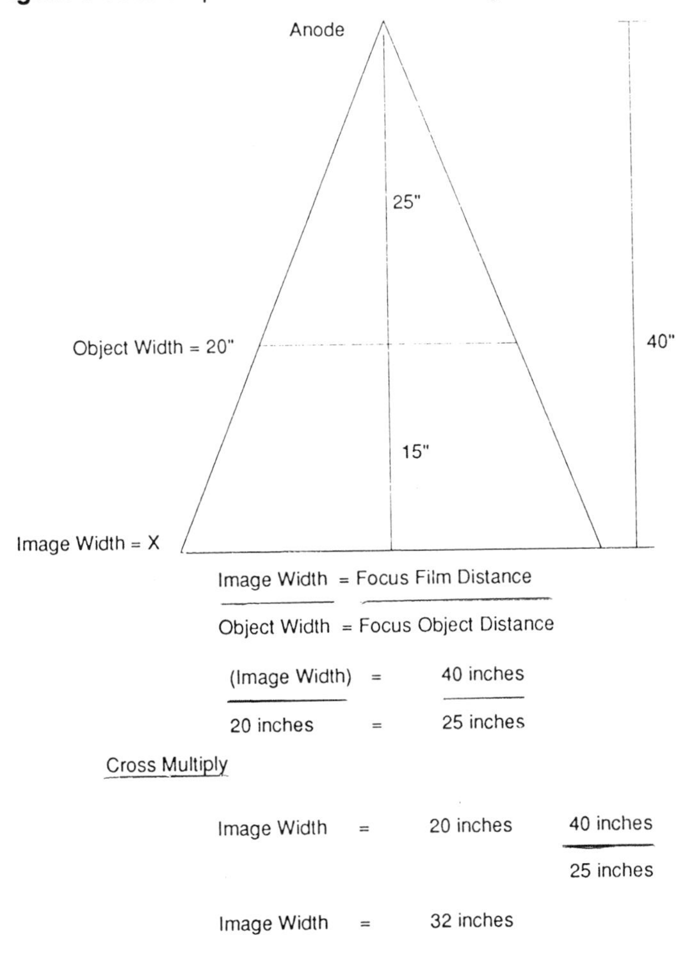

Figure 3-8. Calculation of percentage of magnification.

$$\text{Percentage of Magnification Formula} = \frac{(\text{Image Width} - \text{Object Width}) \times 100}{\text{Object Width}}$$

Example:

$$\text{Percentage of Magnification} = \frac{(32 \text{ inches} - 20 \text{ inches}) \times 100}{20 \text{ inches}}$$

$$X = \frac{12 \text{ inches} \times 100}{20 \text{ inches}}$$

Therefore, $X = 0.6 \times 100$

or, $X = 60\%$

Radiographers control size distortion of the structural details of the radiographic image through the correct manipulation of object-film distance (OFD), focus-to-object distance (FOD), and focus-to-film distance (FFD).The reader may encounter in the literature the abbreviation, TFD. This is the same distance measurement as FFD. However, the abbreviation stands for either electron Target-to-Film Distance, or x-ray Tube-to-Film Distance. Some authors may use the term Source-to-Image Receptor Distance, SID, to describe FFD or TFD. All of these terms (FFD, TFD, or SID) are essentially the same and reflect the measurement of a distance that begins at the level of the focal spot (electron target) and continues to the plane of the x-ray film surface.

Shape Distortion (True Distortion)

Shape distortion is defined as the misrepresentation of the true image through foreshortening or elongation of the structural details of the radiographic image (Figure 3-9). Simply stated,

Figure 3-9. Examples of shape distortion becaues of tube-object-film alignment. (A, aligned; B, distorted due to movement of film plane; C, distorted due to movement of object; D, distorted due to central ray [tube] movement.)

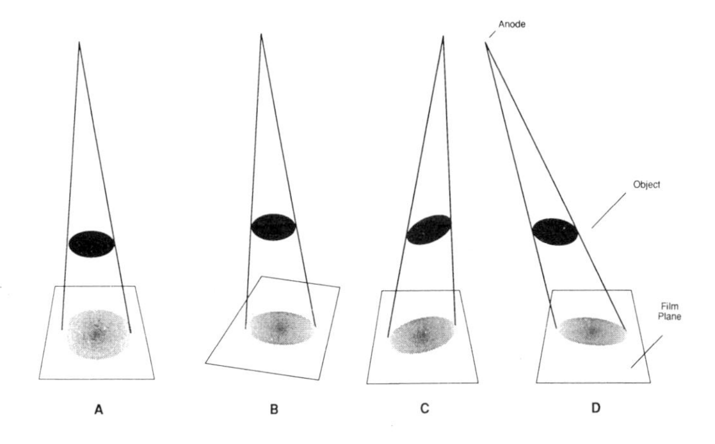

shape distortion will occur if either the tube, the part being radiographed, or the film is angled. This is not necessarily a negative or distracting imaging characteristic. Sometimes, a distorted representation of the object being radiographed is desirable or necessary for diagnostic interpretation. For example, the semi-axial, anterior-posterior "Townes" projection of the skull results in a distorted radiographic image through central ray angulation so that the occipital bone of the skull and other important cranial vault anatomy may be more effectively demonstrated (Figure 3-10A). In a true anterior-posterior view (posterior-anterior projection) of the skull, the occipital bone is distorted (foreshortened) by not being placed in profile with respect to the film plane. In other words, the broad surface of the occipital bone is not parallel to the film. Demonstrating this radiographically in a correct posterior-anterior position is impossible since it is hidden by overlapping cranial structures such as the frontal bone, paranasal sinuses, petrous ridges, and sphenoid bone (Figure 3-10B). Only after performing a special imaging technique is the occipital bone capable of being demonstrated in a proper posterior-anterior anatomical relationship. This requires angulation of the central ray 37 degrees to an infraorbitalmeatal line which is perpendicular to the table top (Figure 3-10A). In order to obtain this elongated view of the occipital bone, however, many gross skull structures are elongated and foreshortened so they may be moved out of the way to assist viewing capability. In this case shape distortion of the gross skull structures is deliberately done by specific guidelines that account for correct angular placement or positioning of the anatomical part in relationship to the central ray of the x-ray tube and film plane. When shape distortion is too extreme and image visualization is disrupted, a repeated radiograph may be warranted.

Figure 3-10. Distortion comparison of two skull radiographs for purpose of viewing the occipital bone. (Notice in "A" [Townes view] that a 37-degree caudal angle of the central ray distorted [elongated] the skull anatomy so that the occipital bone may be clearly visualized. In "B", the occipital bone is obscured by facial bone anatomy and is projected more on end.)

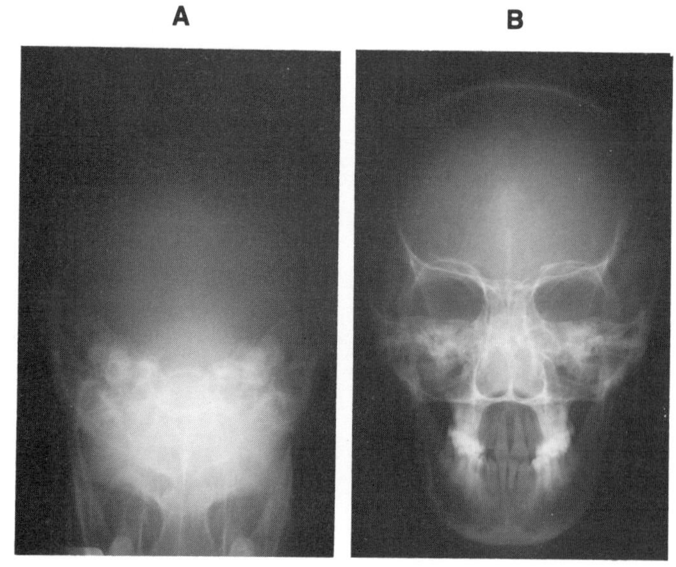

SUMMARY

In addition to creating proper radiographic densities and contrast for visibility of the structural details of the radiographic image, all radiographic images must possess a proper geometric balance. That is, the recorded image details must be sharply defined and of the proper anatomical size and shape. (See Table 3-3 for a summary of the principle factors that affect the geometric properties of image formation.) A radiograph with recorded definition of image details is one that possesses minimal penumbral effect. Penumbra is the area of image unsharpness while umbra is the area of image sharpness. Penumbral effect may be calculated as geometric unsharpness when relationships between focal spot size, OFD, FOD are known. Screen unsharpness is based on the phosphor layer thickness. Faster speed screens usually have thicker phosphor screen layers (except for rare-earth type) which cause more light and a greater light diffusion pattern which increases the unsharpness of the image. This light diffusion pattern may be additionally exaggerated when screens lack close contact with x-ray film surfaces. Motion unsharpness may be the result of involuntary or voluntary patient movement or movement of the x-ray tube or recording medium during exposure. Use of short exposure times with faster screen speeds and x-ray tubes, buckys, and film holders that are properly locked or secured by tape or immobilizing devices helps tremendously to reduce motion of the radiographic image.

Radiographic enlargement of the actual size of the structural details refers to size distortion. Under most situations enlargement of the radiographic image details is not desirable due to the penumbral effect created by either increased OFD or decreased FOD. Macroradiography which employs fractional focal spots is an exception. Aspects of size distortion which may be calculated from the geometric relationships between the focal spot, object, and film plane include the Magnification Factor, Image Width, and Percentage of Magnification.

TABLE 3-3.
Summary of the Principle Factors Affecting the Geometric Properties of Image Formation.

Recorded Detail	Distortion
Geometric Unsharpness	*Size Distortion*
Focal Spot Size	Object Film Distance
Focal Film Distance	Focal Film Distance
Object Film Distance	
	Shape Distortion
Screen Unsharpness	Target, Object, Film Alignment
Film-Screen Contact	
Phosphor Layer Thickness	
Motion Unsharpness	
Tube, Object, Film Movement	

Foreshortening or elongation of the structural details of the radiographic image defines shape distortion. Shape distortion of image details is desired at times to visually demonstrate complicated anatomy on the radiograph. Consider the "Townes" projection of the skull which is designed to correctly demonstrate the occipital bone of the skull which would not normally be visualized in a routine anterior-posterior relationship.

STUDY QUESTIONS

1. What is the area of unsharpness surrounding the radiographic image? What is the area of sharpness?

2. What type of image unsharpness is due to the relationships between focal spot size, OFD, FFD, and FOD?

3. Which of the following sets of information will result in the least amount of geometric unsharpness?

	A.	B.
Focal Spot Size	0.3 mm	0.6 mm
Object Film Distance	2.0 in	4.0 in
Focus Object Distance	38.0 in	36.0 in

4. What type of image unsharpness is the result of poor screen contact?

5. Calculate the magnification factor if the width of the radiographic image is 40 inches and the width of the object being radiographed is 10 inches.

6. Calculate the image width if the object width is 10 inches using a 40-inch FFD and a 30-inch FOD.

7. Calculate the percentage of magnification if the image width is 20 inches and the object width is 5 inches.

8. Shape distortion is to ____________ as size distortion is to ____________.
   A. foreshortening, magnification
   B. elongation, foreshortening
   C. enlargement, magnification
   D. elongation, enlargement
   Answer:
   1. All of the above
   2. A and D
   3. None of the above
   4. B and C

BIBLIOGRAPHY

1. Cahoon, John B., *Formulating X-Ray Techniques*, 8th ed: Duke University Press, Durham, North Carolina, 1974.
2. Donahue, Daniel P., *An Analysis of Radiographic Quality*, Lab Manual and Workbook, University Park Press, Baltimore, Maryland, 1980.
3. Selman, Joseph, *The Fundamentals of X-Ray and Radium Physics*, 5th ed.: Charles C. Thomas, Springfield, Illinois 1976.

Chapter 4

The Perceptual Properties of Image Quality

Chapter Outline

1. Learning Objectives
2. Introduction
3. Radiographic Noise
   A. Artifacts
      1. Sensitized
      2. Nonsensitized
      3. Physical
4. Viewing Conditions
   A. Sight of Viewer
   B. Environmental Conditions
   C. Viewing Distance
   D. Position and Format
5. Evaluating Radiographic Quality
   A. Radiographic Film Evaluation Criteria
   B. Criteria for Repeating a Radiograph
6. Summary
7. Study Questions
8. Bibliography

Learning Objectives

Upon completion of this chapter, the student should be able to:

1. Define radiographic noise, artifacts and mottle.
2. Define sensitized, nonsensitized, and physical artifacts and list three examples of each.
3. List three important conditions for properly viewing radiographs.
4. List three areas from which objective radiographic film evaluation criteria may be developed.
5. State the general criteria for repeating a radiograph.

Introduction

When analyzing the properties that influence the perception of the radiographer for determining the quality of radiographic image details, authors often focus on the intrinsic photo-radiographic and geometric properties of radiographic quality. This is understandable since the intrinsic quality of the radiographic image details significantly impacts on the ability of physicians to conduct an accurate radiological diagnosis of the patient's condition. Consider, however, that there are additional properties of radiographic quality that may influence the way that the radiographic image is perceived. These include radiographic noise, the conditions under which the radiograph is viewed, and the objective criteria radiographers use to evaluate the overall quality of image details.

These additional properties have the potential to significantly effect the viewer's perception of intrinsic radiographic qualities much in the same manner as mAs or kV can affect radiographic density and contrast. Consider the radiographer who does not question the presence of an artifact on a radiograph that might be pathologically perceived by the physician, or consider the radiographer who allows the viewing of radiographs using unevenly illuminated viewboxes in a smoke-filled, noisy work environment; or one who does not base radiographic film evaluation on objective criteria. In all of these instances, the quality of patient care and the quality assurance mission of the entire radiology department may be seriously impaired.

Radiographic Noise

A study of the perceptual properties of image quality would not be adequate without an understanding of the impact that radiographic noise can have on the perception of the image details. Radiographic noise is a term used to describe unwanted radiographic densities on the radiograph that tend to obstruct the viewing of the structural details of the radiographic image. Radiographic noise may be subdivided into two areas, artifacts and mottle.

Artifacts

Anything that was not intentionally meant to be imaged on the radiograph other than the anatomy of the patient may be considered an artifact. Artifacts are considered something extra and contribute to unwanted radiographic densities that tend to distract or obscure the visibility of the structural details of the radiographic image. If gone undetected, some artifacts have the potential for being interpreted as pathology. Thus, artifacts have the potential of interfering with the quality of the radiologic interpretation. Artifacts can originate from a variety of sources including improper handling of x-ray film, static electricity, and fogging. Artifacts may be classified as sensitized, nonsensitized (Kirby, 1975) and physical.

Sensitized artifacts result from any type of exposure energy that is capable of ionizing unexposed silver bromide crystals in the x-ray film emulsion. Commonly encountered sensitized artifacts in medical radiography include static marks and fog. Static marks are a result of electrostatic charges being released on the x-ray film emulsion prior to processing. The type of static artifacts produced depends on the nature of the electrostatic charge. Three common types of static artifacts are tree, crown, and smudge (Figure 4-1). A common method of preventing static artifacts from happening is to electrically ground x-ray film loading benches which will dissipate static electricity accumulated by the radiographer.

Fogging artifacts are the result of electromagnetic or chemical energies exposing x-ray film emulsion. Radiographs that have been fogged have an unwanted density layer throughout all or parts of the processed radiograph. Electromagnetic fogging occurs from two sources: light and radiation. Light fog is the result of visible light reaching the film emulsion prior to processing through light leaks in cassettes, unintended opening of the darkroom door or turning on darkroom white lights (Figure 4-2), or through violation of appropriate darkroom safelight operating conditions. Radiation fog results from inappropriate use of film around x-ray or gamma radiation. Chemical fogging results from allowing the exposed x-ray film to be processed under inappropriate time, temperature, or specific activity of developer or fixer chemistry solutions.

Figure 4-1. Examples of sensitized artifacts; "Tree, Crown, and Smudge" electrostatic artifacts.

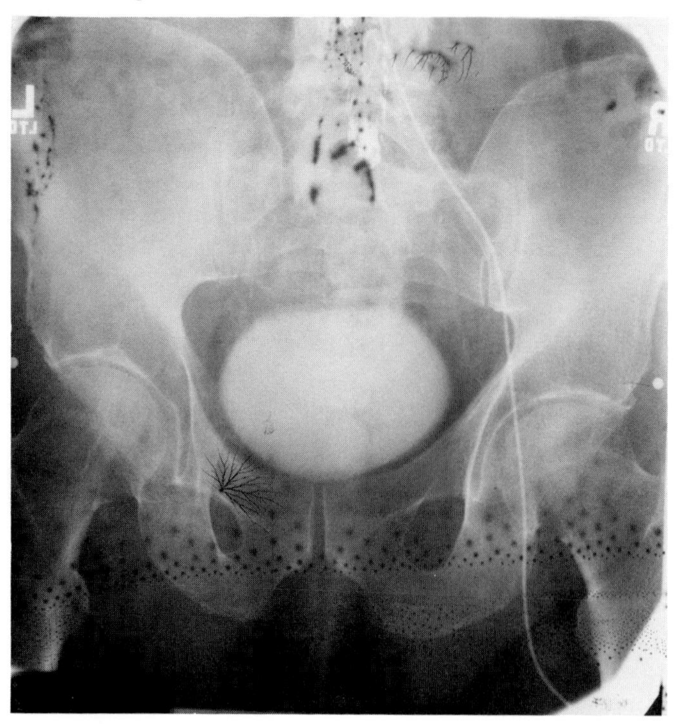

Nonsensitized artifacts result from physical damage of the x-ray film emulsion. Examples of commonly encountered nonsensitized artifacts are pressure artifacts, scratches, spots, streaks and stains. Pressure artifacts result from improper handling of x-ray film as it is being loaded into cassettes from the film bin or from cassettes into the automatic processing unit. Holding x-ray film by the edge and lifting will result in crimp or moon crescent marks (Figure 4-3). Finger prints of the darkroom film handler may be imprinted or superimposed over radiographic image details if fingers are pressed too hard on the soft gelatin film emulsion surface (Figure 4-4). Scratches of the x-ray film emulsion surface can occur before, during or after processing of x-ray film. White spots that appear on radiographs may be caused by unclean intensifying screens (Figure 4-5) or dirt or deposits from an unclean film processor. A condition known as screen lag or after-glow will generate darkened areas on the radiographic image (Figure 4-6). Streaks may be caused by inappropriate agitation of processor chemistries in which there is nonuniformity of development or fixation of the radiographic image details. Residual developer or fixer chemistry that was not adequately rinsed from the film emulsion during processing, can result in a yellow-amber stain being manifested later within the processed film emulsion.

Physical artifacts pertain to objects in, on, or around the patient that have also been made part of the finished radiographic image. Examples of objects *within* a patient are foreign bodies such as bullet fragments, glass chips, metal sutures, radionuclide implants, and internal residual contrast media. Examples of foreign objects that may be found *on* the patient are hair pins, false teeth, buttons, jewelry, necklaces, and rubber bands (Figure 4-7). Examples of foreign objects found *around* the patient are metal traction devices, catheters, monitoring wires, or intubation tubing. (For an in-depth study of radiographic artifacts, the reader is encouraged to consult the textbook by Richard J. Sweeney entitled, *Radiographic Artifacts: Their Cause and Control*.)

Mottle

Radiographic mottle is defined as the density variation in a radiograph made with intensifying screens which have been given a uniform x-ray exposure (Kodak, 1980). Mottle consists of quantum mottle, structure mottle and film graininess. This discussion is limited to scientific definitions and will not be considered in depth since the disruptive aspects of mottle and film graininess are minimal except in the case when rare earth type intensifying screens are used during extremely short exposure times.

Quantum mottle is defined as the variation in density of a uniformly exposed radiograph that results from the random spatial distribution of x-ray quanta absorbed in the screen (Kodak, 1980). Given a uniformly structured intensifying screen that gives a uniform light exposure and x-ray quantity that is not uniformly distributed across the intensifying screen (due to poor statistics), one is likely to view quantum mottle on the radiograph. Quantum mottle may be affected by film speed, film contrast, screen conversion efficiency, screen absorption, radiation quality and light diffusion (Kodak, 1980).

Structure mottle is defined as the density fluctuation resulting from nonuniformity in structure of intensifying screens used to make the exposure. Such nonuniformity in structure may be attributed to nonuniformity in the phosphor layer, such as clumping of the phosphor crystals, or coating variations. The Combined effect of structure mottle and quantum mottle is sometimes referred to a screen mottle. However, in most commercially available screens, structure mottle is not a problem (Kodak, 1980).

Film graininess is defined as the visual impression of the density variation in a film uniformly exposed to light (not light from intensifying screens, because exposure by intensifying screens produces quantum mottle which overwhelms film graininess).

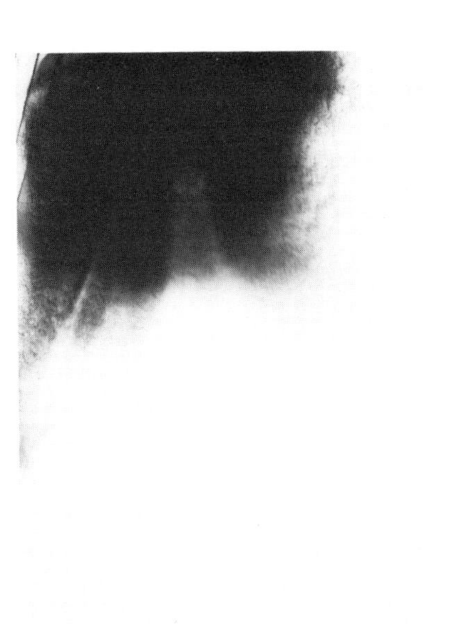

Figure 4-2. Example of "light fogging" as a result of a pinhole light leak in a direct exposure holder.

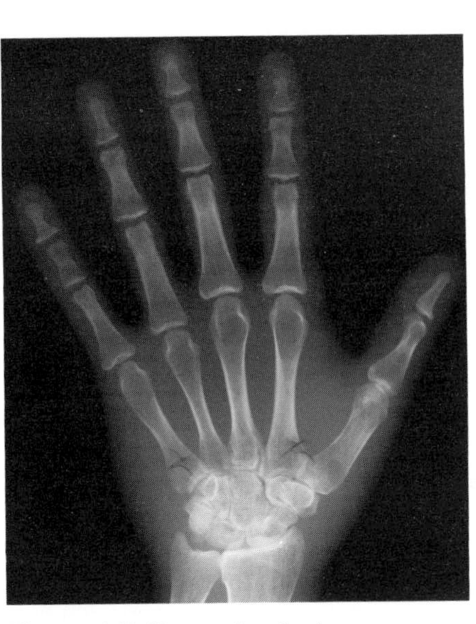

Figure 4-3. Example of crimp or moon crescent artifacts.

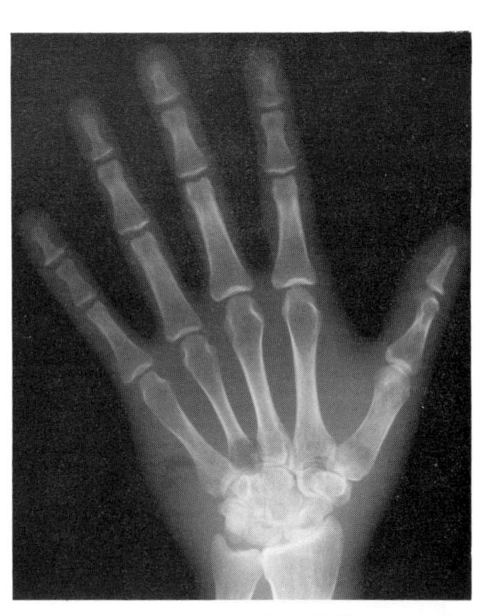

Figure 4-4. Example of finger-print artifacts.

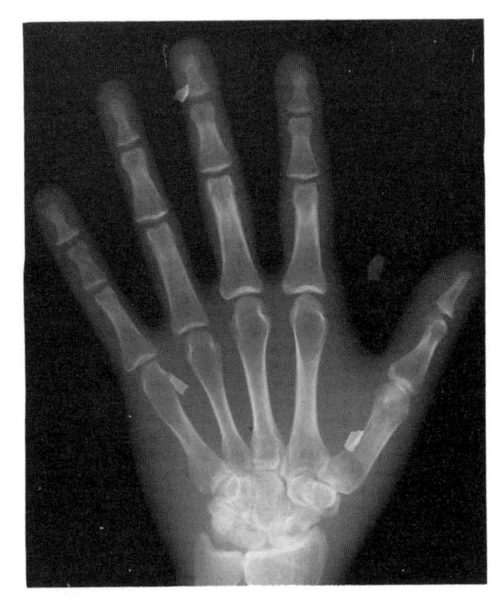

Figure 4-5. Example of "white spot" artifacts resulting from unclean intensifying screen-cassettes. (Notice how the chips of paper that are in the cassette resemble glass or metallic fragments that could be misinterpreted by the attending physician.)

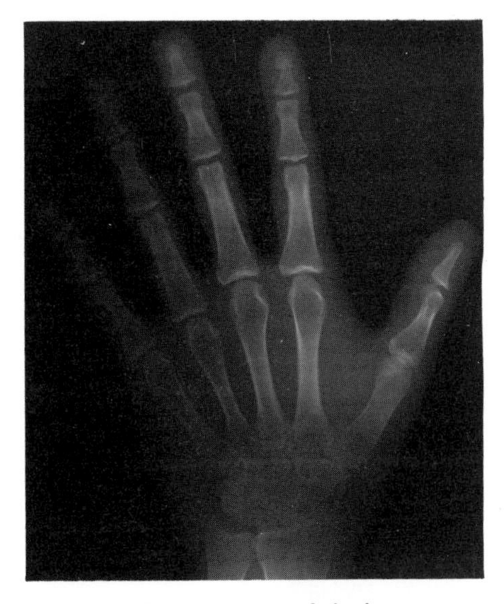

Figure 4-6. Example of dark area artifacts resulting from screen lag (afterglow).

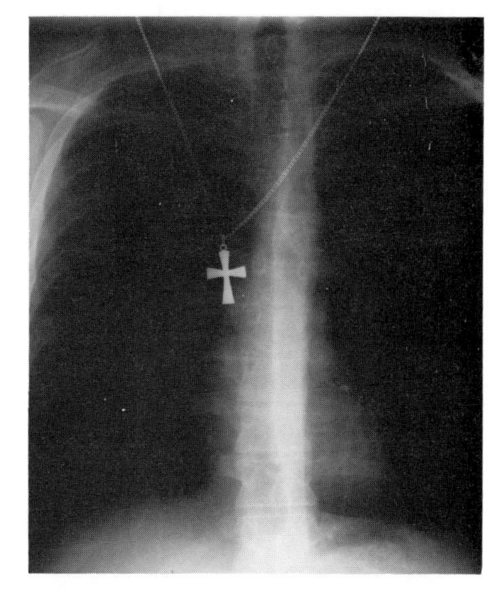

Figure 4-7. Example of a physical artifact located on the patient.

Film graininess is due to the random distribution of the deposits of developed silver in the x-ray film emulsion (Kodak, 1980).

Viewing Conditions

The radiographer is usually the first to view and evaluate the quality of the finished radiograph before it is given to the physician for interpretation of the patient's condition. Usually, the first thing that the radiographer may do after determining if density, contrast, recorded definition of detail, and distortion are acceptable is to make sure that all anatomy is correctly displayed according to proper positioning landmarks and baselines. Obliques or rotated positions are checked along with the correct alignment of the central ray of the x-ray beam to the anatomical part which in turn should be correctly aligned with the center of the film. Finally, after aspects of radiation protection, patient identification, and film markers are checked, the intrinsic qualities of the radiograph are either judged acceptable for interpretation by the physician or they are not. If not, a repeated film may be required to correct any particular imaging problems.

Often overlooked by the radiographer is the impact on image quality because of viewing conditions. When viewing conditions are less than optimal, quality radiographs may actually appear to be lacking in quality and this could negatively affect the quality of the eventual diagnosis. Four important factors that may affect the viewing of radiographs for film evaluation or diagnosis include the sight of the viewer, viewing environment, viewing distance, and viewing format.

The sight of the viewer is rarely considered as having the potential of impacting negatively on viewing radiographs. However, consider the impact on viewing if lenses in glasses are dirty, prescription lenses are outdated, or contact lenses and glasses fit poorly. Individuals who are far sighted may require corrective lenses as radiographic viewing distances are relatively short. Most individuals with a vision problem usually seek immediate medical attention and have it corrected. However, some may not be aware they have a problem while others may be too vain to wear corrective lenses. As the work day comes to an end, radiographers need to be aware that their eyes may be stressed and tired; especially after the rigors of frequently having one's eyes adjusting to darkroom and viewbox lighting conditions.

The environmental conditions for viewing must also be optimal. For example, viewbox illuminators must be free of dirt or dust. Illuminators should project a uniform fluorescent white light intensity, especially when more than one viewbox illuminator is used so that false perceptions in contrast and density are prevented. Likewise, there shouldn't be any glare in the viewing room from extraneous light, other illuminators or lights external to the viewing room. The ability to concentrate during film evaluation or diagnosis is enhanced if distracting noises and movements are kept to a minimum in the immediate viewing en-

Figure 4-8. Example of (A) correct placement and (B) incorrect placement of PA chest radiographs on the viewbox. (Notice in (A) correct placement of radiograph that the patient's heart shadow is facing the viewer's right side. This is an indication that the patient is facing viewer in the anatomical position.)

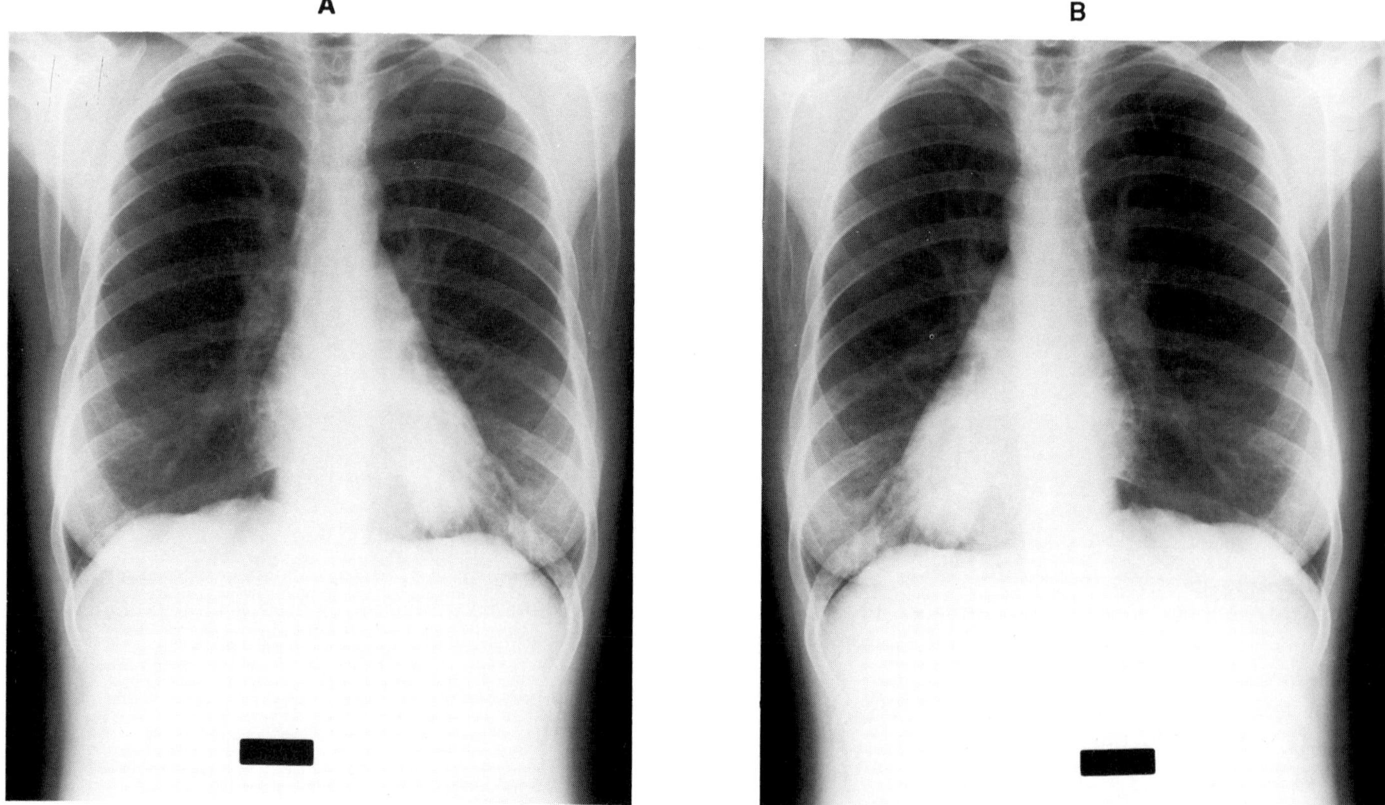

vironment. Smoking should be discouraged while viewing radiographs since smoke irritates the eyes and has the potential of obscuring image details.

The eye views a radiograph as a collection of points. If these points are not less than one one-hundredths (1/100ths) of an inch in diameter and the viewing distance is ten inches, all points appear sharp (Jacobi and Paris, 1977). Therefore, radiographers should place themselves at a relatively short radiographic viewing distance. For all practical purposes, viewing distances beyond one arm's length might jeopardize visual acuity.

Radiographs should be viewed in the proper anatomical image display format. Assuming that the radiographer has correctly marked the right and left side of the body (with additional direction or time markers if required), all radiographs should be displayed such that the image is as if the patient is in the anatomical position. That is, the patient's right side is on the viewer's left side. For example, when placing an A.P. view of the chest (P.A. projection) on the viewbox, the thorax should appear as if the patient is facing the viewer (Figure 4-8).

In order to make viewing of radiographs less complicated they should be organized in a proper format. For example, with multiple images on one radiograph as with the wrist, anatomy such as fingers and forearms should be placed in the same orientation on the film for all exposures in order to simplify viewing and interpretation. Radiographers and physicians should not be required to turn their heads at extreme angles or turn radiographs up or down to correctly orient the anatomy. Likewise, where two or more radiographic studies are performed on a patient, the resulting radiographs should be presented on the viewboxes in terms of their separate examination category and not as an integrated synthesis of work performed.

Evaluating Radiographic Quality

Radiographers should learn how to objectively evaluate the quality of a radiograph based on an organized evaluation strategy that takes into account the essential factors that may impact on the diagnostic acceptability of the radiograph. This type of evaluation should not be confused with the ability to diagnose from radiographs which the physician is qualified to perform (Lauer, 1984). The radiographic evaluation process may be considered another perceptual factor that could influence the interpretation of radiographic quality by a radiographer since radiographers who lack an objective approach to analysis of radiographic quality are more likely to be subjective and less critical in correctly judging the diagnostic acceptability of radiographic images.

The finished radiograph is the product of many tasks that went into its construction. For example, consider that from viewing a radiograph the viewer, acting like a detective, may be able to determine the approximate nature of the affecting factors responsible for the quality of image visibility, formation, and image quality. For example, with an overexposed radiograph, in which image details are obstructed by too much radiographic density, one may be able to deduce if one or more of the principle affecting variables of mA, time, kV, or FFD, may be at fault. If there is unexpected distortion, alignment of tube, part, or film is at fault. Once the radiograph is on the viewbox, the perception of image quality may be overshadowed by uneven illumination, artifacts, or extraneous light glare.

Radiographic Film Evaluation Criteria

Since the radiograph represents many known imaging events that went into its construction, developing criteria for radiographic film evaluation based upon these events provides a prudent basis for objective analysis of radiographic quality. The following radiographic film evaluation criteria is provided as a guide for the objective analysis of radiographic quality.

A. Criteria for Evaluating Photo-Radiographic Properties
   1. The structural details of the external and internal radiographic image must be visible so the attending physician can correctly identify the anatomy of interest.
      a. Are densities sufficient for image construction?
      b. Is there sufficient contrast to distinguish between differences in tissue composition?
      c. Were the principle exposure factors of mA, Time, kV, and FFD correctly employed in relationship to the anatomy radiographed and the recording system?
      d. Were the film-processing parameters of time, temperature, and specific activity of development correct?
      e. Were the appropriate grids and beam restrictors used to prevent "fog"?

B. Criteria for Evaluating Geometric Properties
   1. The structural details of the radiographic image must be sharply defined and of the correct size and shape for the radiographic procedure in question.
      a. Are the structural lines of the image details clearly recorded?
         (1) Was the correct focal spot size, FFD, and OFD employed?
         (2) Was the correct film/screen combination used with excellent film screen contact?
         (3) Was there no movement between the x-ray tube, part radiographed, and film during exposure?
      b. Does the shape (broad surface and long axis) of the image represent the correct anatomy relative to the radiographic projection performed?
         (1) Was the x-ray tube, part radiographed, and film correctly aligned during exposure?

C. Criteria for Evaluating Perceptual Properties
   1. There should not be any radiographic noise within the image to obstruct visibility or image form.
      a. Are any sensitized, nonsensitized, or physical artifacts present that obstruct visualization of anatomy or question the presence of pathology?
      b. Is quantum mottle, structural mottle, or film graininess present and sufficient to obscure visualization?
      c. When two or more radiographic views of an anatomical part are present on a single radiograph (or in more than

two separate radiographs), do they all run in the same anatomical orientation?

2. The radiographic image should reflect the desired anatomy correctly positioned.
   a. Is the anatomy correct?
   b. Are proper baselines and landmarks visible?
   c. Were oblique or rotated projections done correctly?
   d. Was the central ray of the x-ray beam correctly aligned to the anatomical part and the center of the film?
   e. Does the film size used for the part radiographed reflect departmental standards) (If deviations in film size were made, are mitigating circumstances warranted?)
   f. Was the film placed in the correct longitudinal, transverse, or erect position. (Note: Even in erect film planes, films must be either in a longitudinal or transverse placement.)
3. The radiograph must be identified in relationship to the patient and radiographer, and it must be correctly marked.
   a. Is the appropriate patient identification present?
   b. Are the radiographer's initials present?
   c. Are the correct directional and informational markers present so as not to obscure the visualization of the desired anatomy?
4. There should be evidence of radiation protection of the patient on the radiograph.
   a. Is there evidence that beam-restricting devices were used?
   b. When applicable, is there evidence of part or gonadal shielding?
   c. Were radiographic exposure techniques (mAs, kV, and FFD, screens) employed that are designed to minimize radiation dose to both patient and operator?
5. The quality of the radiographic image should be assessed using objective criteria (such as that now being identified).

Criteria for Repeating a Radiograph

Radiographers are faced at times with the important responsibility of deciding whether or not to repeat a radiograph. This responsibility is important because the act of repeating a radiograph causes additional radiation dosage to the patient (and possibly radiographer) and results in additional resource utilization. Therefore, the act of repeating a radiograph is both hazardous and expensive. By what criteria should a radiograph be repeated? Certainly, a radiograph should not be repeated simply because, for example, there was no evidence of gonadal shielding, the radiograph was of the wrong film size, it was incorrectly marked, or views were not oriented properly. Hopefully common sense will dictate that as long as the desired structural details of the radiographic image are visible and sharply defined, repeated radiographs are not necessary, regardless of any lack of aesthetic quality. On the other hand, the following situations represent some justified reasons why a radiograph may need to be repeated:

1. Excessively blurred image details due to patient motion or excessive size distortion.
2. Failure of the x-ray machine to deliver the desired quantity or quality of radiation to the film due to electronic failure.
3. Failure of the radiographer to place the tube directly over the part to be radiographed in a locked position.
4. Failure of the processor to adequately develop and fix the radiographic image based on optimal time, temperature, and specific activity of development.
5. Excessive overexposure or underexposure because the radiographer failed to follow or use valid reliable radiographic exposure technical guidelines.
6. When the shape of image details are so excessively misrepresented that the anatomy of interest is not recognizable.
7. When the anatomy is obscured from view due to positioning errors.
8. When an artifact is questioned as possible pathology.

SUMMARY

Aside from the intrinsic photo-radiographic and geometric qualities of the radiographic image, radiographers should consider that there also exists extrinsic aspects known as perceptual properties of image quality. These perceptual properties may affect the viewer's perception of the photoradiographic and geometric qualities. Viewer perception may be influenced by radiographic noise, viewing conditions, and the criteria by which the radiographic images are assessed for diagnostic acceptability. Radiographic noise pertains to unwanted radiographic densities generated from sensitized, nonsensitized, and physical artifacts and quantum mottle, structural mottle, and film graininess. Factors that may affect radiographic viewing conditions include visual acuity of the viewer, conditions of the viewbox illuminators, extraneous light, noise, smoking, and viewing distance.

The criteria for objective analysis of the quality of radiographic image details should include analysis of the photoradiographic, geometric, and perceptual properties of radiographic quality. Assessment of radiographic positioning and radiation protection should be included. Hopefully common sense will dictate that as long as the structural details of the radiographic image are visible and sharply defined, repeated radiographs are usually unnecessary, regardless of any lack of aesthetic quality. Some situations may warrant repeated radiographs. However, the implementation of a quality assurance program based on analysis of retakes and quality control testing should assist in preventing retakes even

TABLE 4-1.
Summary of the Principle Factors Affecting the Perceptual Properties of Image Quality.

Radiographic Noise	View Conditions	Evaluation Criteria
Artifacts	Sight	Photo-Radiographic
Sensitized	Environment	Geometric
Non-Sensitized	Distance	Perceptual
Physical	Display Format	
Mottle		
Quantum		
Structural		
Film Graininess		

with respect to these seemingly justified situations. Table 4-1 provides a summary of the principle factors affecting the perceptual properties of image quality.

STUDY QUESTIONS

1. List three factors that may influence the perception of the diagnostic acceptability of a radiograph.
2. Define radiographic noise.
3. Define artifact.
4. Define sensitized, nonsensitized, and physical artifact, and give two examples of each.
5. List three conditions that may impact negatively when viewing radiographs.
6. List several factors that affect the visibility of image details of the radiographic image.
7. List several factors that affect the geometry of image formation of the radiographic image.
8. List several factors that affect the anatomic position of the radiographic image.
9. List several factors that indicate evidence of radiation protection on the radiographic image.
10. State the general criteria for repeating a radiograph?
11. Give three examples of situations that seemingly justify the repeating of a radiograph.

BIBLIOGRAPHY

1. Jacobi, Charles A., and Don Q. Paris, *Textbook of Radiologic Technology*, 6th ed.: C. V. Mosby, St. Louis, Mo., 1977.
2. Kirby, Cynthia, C., William E. J. McKinney, and Thomas T. Thompson, *A Guide for Automatic Radiographic Processing and Film Quality Control*, The American Society of Radiologic Technologists, Chicago, Illinois, 1975.
3. Lauer, O. Gary, "A Method for Radiographic Film Evaluation," in Lauer, O.G. (ed): *Principles and Practices of the College Based Radiography Program; Educational and Administrative Considerations*, Warren H. Green, Inc., St. Louis, Mo., 1984.
3. Stevens, Mathew and Robert I. Phillips, *Comprehensive Review for Radiologic Technologists*, C. V. Mosby, St. Louis, Mo., 1977.
4. *The Fundamentals of Radiography*, 12th ed.: Eastman Kodak Company, Rochester, New York, 1980.
5. Sweeny, Richard J., *Radiographic Artifacts: Their Cause and Control*, J. B. Lippincott Co., Philadelphia, PA 1983.

Plate II

The Effects Of Principal Exposure And Imaging Factors On Radiographic Quality

	Photo-Radiographic and Geometric Properties				
Principal Factors	Density	Contrast	Recorded Detail	Size Distortion	Shape Distortion
kVp	+	–	0	0	0
mAs	+	0	0	0	0
Distance TFD	–	0	+	–	0
OFD	–	+	–	+	0
Focal Spot Size	0	0	–	0	0
Filtration	–	–	0	0	0
Beam Restrictors	+	–	0	0	0
Tissue Quality	–	+	0	0	0
Grids	–	+	0	0	0
Intensifying Screens	+	+	–	0	0
Motion	0	0	–	0	0
Safelight Fog	+	–	0	0	0
Processing (Time Temperature Specific Activity)	+	+, (–)*	0	0	0

= Direct Relationship
= Inverse Relationship
= No Effect

*When processing conditions deviate from manufacturer's recommended guidelines.

Chapter 5

The Effects of Kilovoltage on Radiographic Quality

Chapter Outline

1. Learning Objectives
2. Introduction
3. The Role of Kilovoltage in Radiography
4. The Effects of kVp on Image Visibility
   A. Radiographic Density
   B. Radiographic Contrast
   C. Exposure Latitude
5. The Effects of kVp on Image Formation
6. Summary
7. Study Questions
8. Bibliography

Learning Objectives

Upon completion of this chapter, the student should be ble to:

Describe the role of kVp within the x-ray tube with regard to the x-rays produced.

Explain the "rule of 10."

Identify the effect, if any, on the following, if there is a change in kVp:
a. Radiographic density
b. Radiographic contrast
c. Radiographic scale of contrast
d. Exposure latitude
e. Recorded detail
f. Size distortion
g. Shape distortion

Given different kVp settings, identify the one that would produce a radiograph with more (or less) density, higher (or lower) contrast, longer (or shorter) scale of contrast, wider (or more narrow) latitude of exposure, as specified.

troduction

Almost every student radiographer has learned during the first onth that kilovoltage affects the penetrating ability or power the x-ray beam. Beyond this many radiographers fail to comehend the basic concept of kilovoltage and its effect on diographic quality. The selection of kilovoltage is so critical that many "repeat" radiographs are taken simply because not all radiographers are able to perceive what will happen to the film as they alter the kVp control. Many are unaware that kVp is even a factor that needs to be changed.

The Role of Kilovoltage in Radiography

KVp is the abbreviation for kilovoltage peak, with voltage being the push and pull on electrons, and kilo being the scientific abbreviation for 1000. (85 kV, for example, could be expressed as 85,000 volts, and 56,000 volts could also be expressed as 56 kilovolts.) The force of regular voltage would be insufficient to drive electrons from the cathode to anode with enough energy to produce radiation suitable for penetrating human tissue; hence kilovoltage is used to impart enough energy for that purpose. The "p" for peak designates the maximum voltage in each burst of electron energy produced in the pulsations of current within the x-ray tube. If a setting of 75 kVp is used on the x-ray machine (75 kVp means that each burst *peaks* at 75 kV), the voltage is less leading up to and going away from that peak.

A review of the role of kV within the x-ray tube may also prove beneficial. As kVp is increased, the electrons which have been produced from the tube filament are accelerated to greater speeds across the tube. As a result, the x-rays that are produced in the focal spot of the anode will have a shorter wavelength, a higher frequency, and will be capable of deeper penetration (Figure 5-1).

Figure 5-1. Different waveforms as affected by kilovoltage.

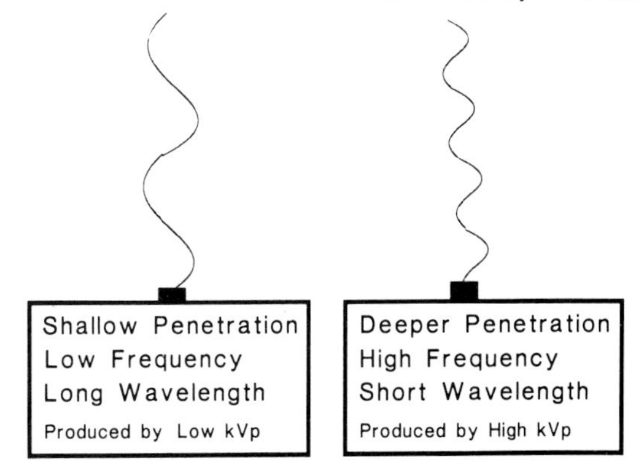

Also the quantity of x-rays produced is affected by a change in kVp. Not all of the electrons which are accelerated across the tube will produce diagnostic x-rays. As the same number of electrons is forced more quickly across the tube (higher kV), a slightly greater percentage will fall within the diagnostic range which means that kVp influences both wavelength and quantity. A higher kVp will produce more x-rays and each will be capable of deeper penetration. Lower kVp will produce fewer x-rays and each will have less penetration.

The intent of this chapter is to clarify not only the theory of what happens within an x-ray tube with respect to voltage, but to show the effects on a film of any voltage changes. Through exercises and examples, radiographers should be provided with the necessary information to comprehend the results of voltage changes and to evaluate results for desirability.

The Effects of kVp on Image Visibility

Radiographic Density

Kilovoltage definitely affects density. Remember, kVp affects both quality (penetrating power) and quantity, in that order. Given a film of the same body part with different kVp settings, where there is greater penetration, more x-rays will reach the film. More x-rays reaching the film will cause the radiographic image to be darker, that is, the image will have more radiographic density. It is understood, then, that kVp affects radiographic density because if kVp is increased, then radiographic density increases. The reverse is also true: if kVp is decreased, then radiographic density also decreases.

The following experiment is designed to assess the hypothesis that kVp affects density. The technical factors that are manipulated on the x-ray unit will be shown for each film so that one can set up a similar experiment. It should be understood, however, that some modifications in these factors may be necessary, depending on experimental conditions such as a type of x-ray film, filmholder, or machine calibration, to name a few. (See Figure 5-2.)

Obviously, kVp does have a direct relationship to radiographic density and does not require a very large adjustment to produce a significant effect on radiographic density. In fact, it only takes a 4 percent change in kVp to create an observable effect. Four percent of a starting kVp of 50 translates into 2 kVp. Four percent of 75 kVp equals 3 kVp. Four percent of 100 kVp equals 4 kVp. Referring again to the set of hand radiographs in Figure 5-2, there are notable differences in radiographic density. If only the slightest density difference were desired, the kVp would need to be changed by at least 2 kVp (from 50 to 52 for slightly more, or from 50 to 48 for slightly less density). The use of 51 kVp or 49 kVp would result in too small an effect to be observable (Figure 5-3).

Normally, kVp is not the primary factor to be manipulated when density is to be adjusted. If kVp were to be used to adjust radiographic density by a large degree (a doubling or halving of the density), then a 15 percent change in the original kVp would be required. For example, from 50 kVp, if a doubling density effect were required, an increase from 50 kVp to 57.5 kVp (or 58) kVp would be needed (15% of 50 = 50 + 7.5 = 57.5). Since most x-ray machines do not generally have the capability to select fractional kVp values, 58 is used as a close approximation.

An approximation of the above principle learned by many radiographers is referred to as the "rule of 10." Simply explained the "rule of 10," which applies in mid-range kVp values, says that each change of 10 kVp will double (or halve) density, depending on the direction in which kVp is adjusted. For example, changing from 60 to 70 kVp will double density. Changing from 90 to 80 kVp will halve density.

In the set of films shown in Figure 5-4, the one in the center will be the control film. The films to the left get lighter in "halving" amounts (minus 15 percent for each film), and the films to the right get darker in doubling amounts (plus 15 percent for each film). Note that penetration and image contrast are significantly affected through large kVp changes.

Fog, an additional density on a radiograph, is usually caused by unwanted scatter radiation reaching the x-ray film, and scatter radiation is also affected by kVp. Since a more penetrating primary x-ray beam causes scatter radiation that is more penetrating, higher kVp will generally produce radiographs with more fog. Devices such as grids and beam restrictors are designed to limit the amount of scatter radiation reaching the film. Simply stated, more fog equals more density; less fog equals less density. Figure 5-5 shows a simple chart relating changes in kVp to the production of scatter radiation, fog, and radiographic density.

Radiographic Contrast

One of the most difficult relationships for a beginning radiographer to understand is contrast. This difficulty may occur

Figure 5-2. Density affected by kVp.

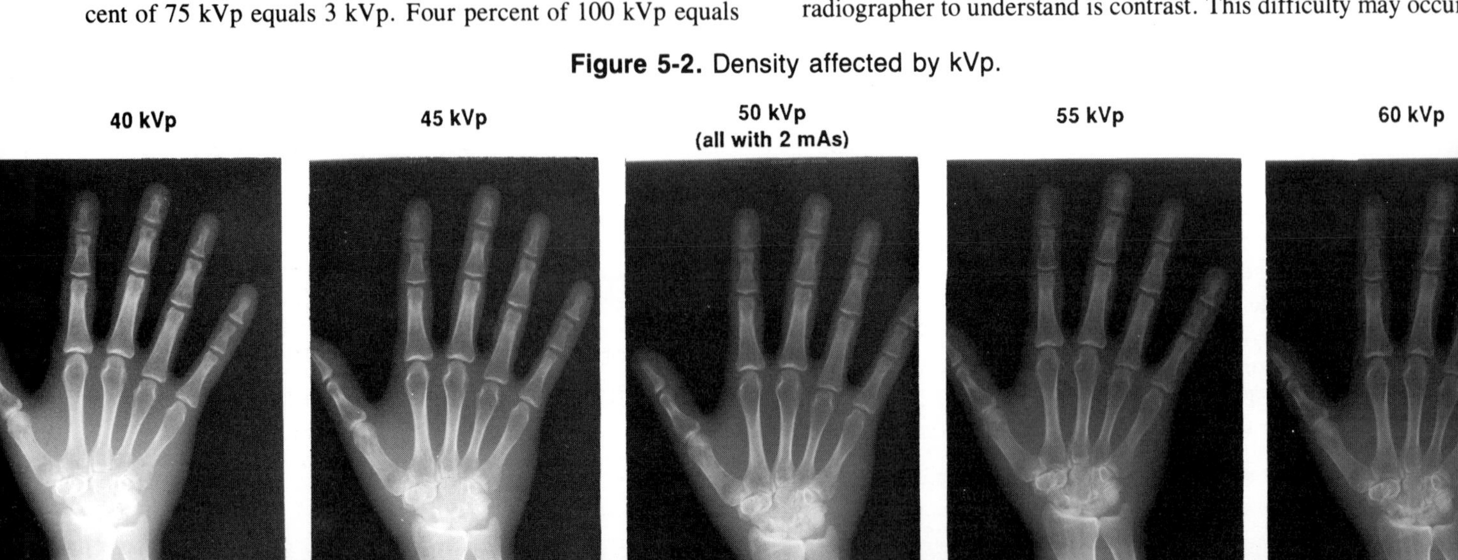

r several reasons. Contrast is not always referred to with the me terminology. Some radiographers may refer to a radiograph ''high contrast,'' while someone else may have learned to call e same radiograph a ''short scale'' film. Part of the reason for is confusion is a lack of standardization of terminology among hools and professionals. Occasionally, density and contrast are nfused. Radiographers may perceive that a radiograph needs change in density, when in actuality, it is contrast that requires ljustment.

Remember that *contrast* means *difference*. Radiographers sociate contrast with the difference between varying shades of ay on the radiograph. Subject contrast and film contrast are e two components of radiographic contrast. For a more in-depth scussion of both components refer to Chapter 2. Suffice it to y, however, that radiographers change the quality and quanti- of the x-ray beam by their selection of machine factors which es alter the radiographic contrast (Figure 5-6).

Thinner, less dense tissues are imaged on x-ray film in darker ades of gray than are the heavier, thicker tissues. The bones a hand radiograph, for example, appear relatively light while e surrounding softer tissues are dark. The bones have stopped ore x-rays and the film emulsion was not as heavily bombarded y x-rays in those areas. The x-ray beam is heterogeneous, that , even though the x-ray machine is set to a given kV, not all x-rays within the beam are exactly the same. Some have slightly more penetrating power, others will have less. The range or spread of this heterogeneity increases with kVp.

You may note from Figure 5-6 that the ''high'' contrast film appears to be more eye appealing. This is often the case; however, one cannot assume that this is a radiograph of the most diagnostic quality. The type of exam, area of interest (bone versus soft tissue; abdomen versus lumbar spine), and type of pathology must dictate the type of contrast to be evident on the final film. Radiographers should remember, however, that low contrast radiographs have more shades of gray. Since each shade represents a particular type of tissue, more information may be revealed to the radiologist when compared with radiographs that have high subject contrast characteristics.

Contrast can be referred to in other ways, but, the terms ''high''

Figure 5-5. Relationships between kVp and Scatter Radiation, Fog, and Radiographic Density.

kVp	Scatter Radiation	Fog	Radiographic Density
+	+	+	+
-	-	-	-

Figure 5-3. Small density adjustments with kVp—4% change per radiograph. (High Speed Screens)

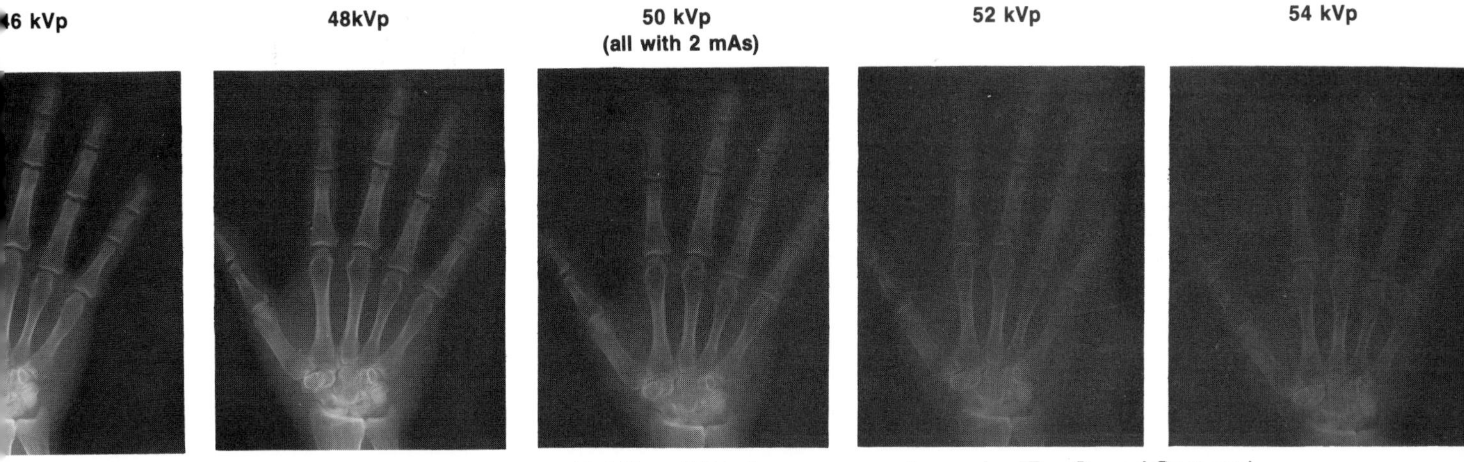

Figure 5-4. Large density adjustments with kVp—15% change per radiograph. (Par Speed Screens)

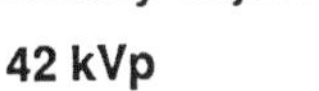

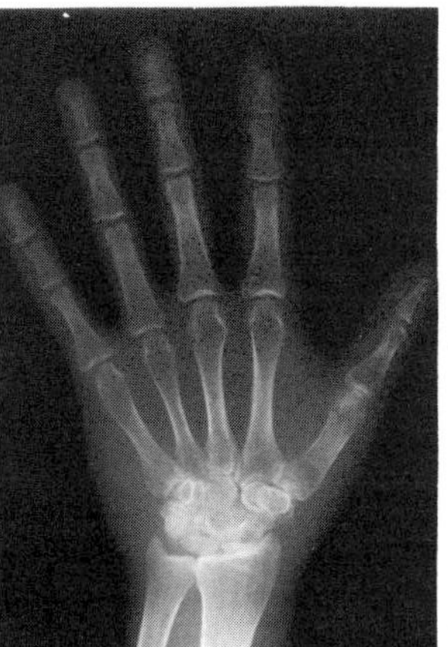
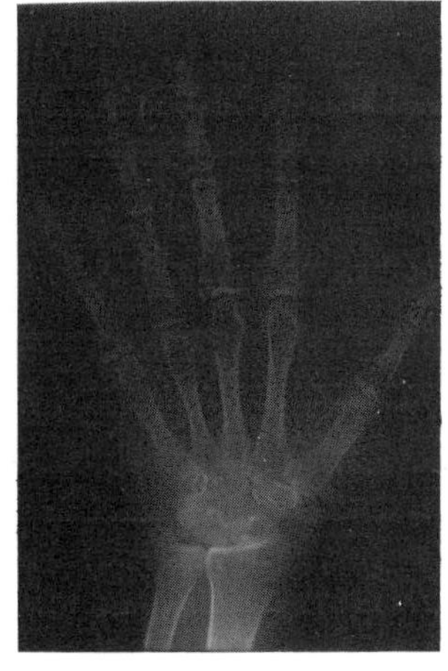

Figure 5-6. Radiographs demonstrating different subject contrast as a result of high and low kVp.

Low kVp
High Contrast
(Short Scale of Contrast)
65 kVp, 40 mAs

High kVp
Low Contrast
(Long Scale of Contrast)
80 kVp, 13 mAs

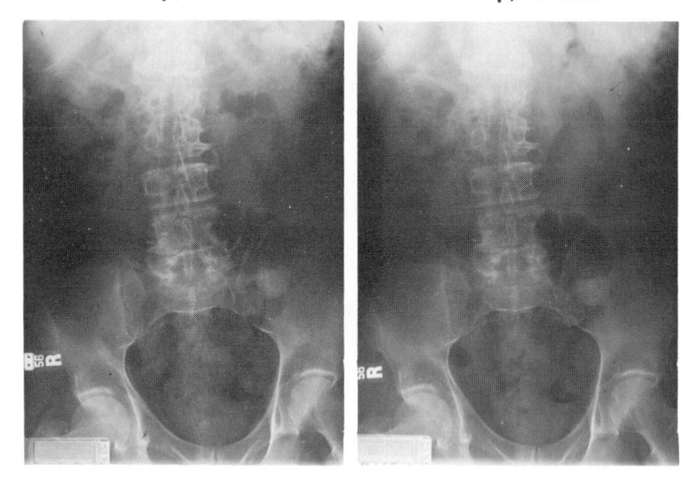

and "low" will be used here to discuss the manipulation of kVp to achieve the type of contrast needed on radiographs. Figure 5-6, illustrates that low kVp gives high contrast, and high kVp gives low contrast. This is called an inverse relationship. Inverse means opposite; hence, subject contrast is inversely related to kVp.

Higher contrast can be obtained by lowering kVp, but this adjustment in kVp will result in a simultaneous decrease in radiographic density. If the goal of a radiographer is to change only contrast, then density should not be altered. This is where compensating factors become important. If a radiographer wanted to lower kVp in order to increase contrast, an increase in mAs would be required to maintain equivalent densities. As far as density is concerned, a 15 percent change would double or halve radiographic density. The radiographer would also need to know the quantity of mAs that would halve or double to offset kVp's density effect. This calculation isn't difficult. For higher contrast, lower kVp by 15 percent; this halves density. To get the density back to its original level, double the mAs. For lower contrast, raise kVp by 15 percent; this doubles density. To get the density back to its original level, halve the mAs (Figure 5-7).

Another point about the preceding films is that anyone can note individual areas on the high contrast film that have more density than the same areas on the other two films, and from that deduce that density has not been balanced. On that same film, however, one could also find areas that are lighter. What that really shows is different contrast. There is more separation between the light and dark parts of the film. Simultaneously, similar tissues having slightly different tones on a longer scale of contrast film are now combined in a single shade and could not be distinguished on the high contrast film. *Remember, the radiograph that is most eye appealing may not necessarily be the most diagnostic.*

While kVp's effect on contrast is inverse, the relationship to the scale of contrast is direct. Scale of contrast refers to the quantity of shades of gray within the density spectrum. Because the use of a low kVp makes the beam more homogeneous, anatomic tissues of similar density would appear to have the same radiographic density. With a higher kVp each tissue might have its own radiographic density. Higher kVp then produces a film with a longer scale of contrast; lower kVp produces a film with a shorter scale of contrast.

Increasing kVp (and thereby lowering contrast or lengthening the scale of contrast) will only work within appropriate ranges of kVp. There is a point beyond which the radiographic contrast is too low and differentiation between structures is so difficult

Figure 5-7. Radiographs demonstrating different contrast and same density.

Low Contrast
92 kVp, 1 mAs

Control
80 kVp, 2 mAs

High Contrast
68 kVp, 4 mAs

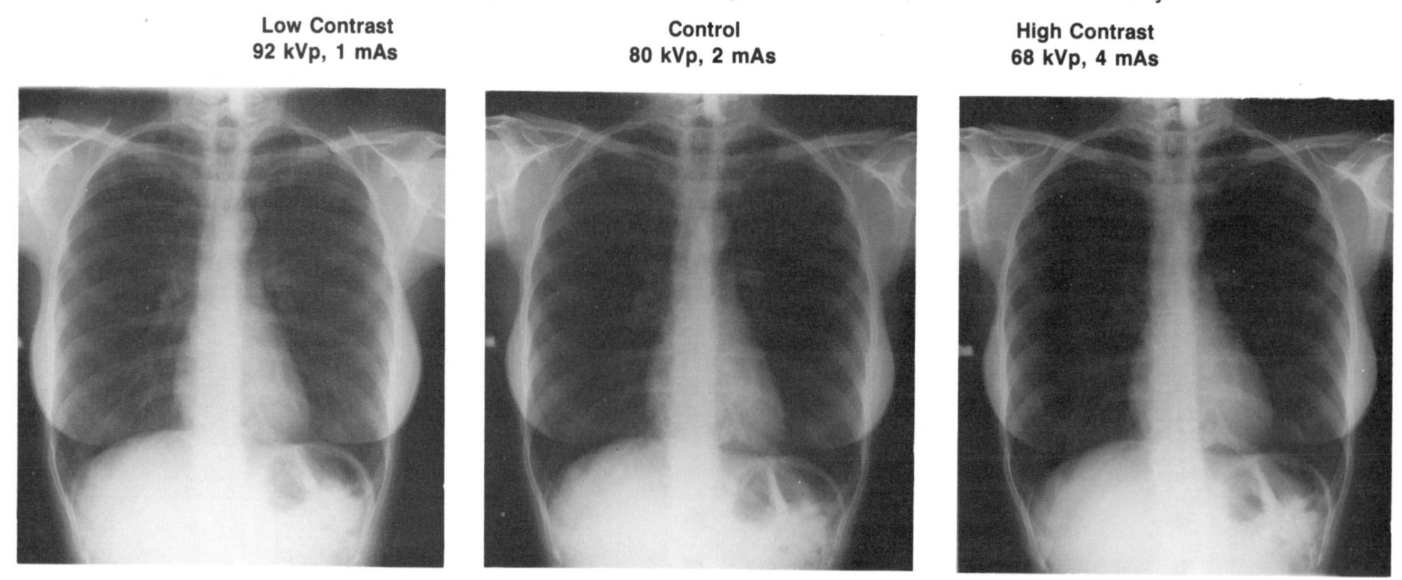

that the film is rendered undiagnostic. Additionally, large changes in kVp could alter tissue penetration to the point where an image is so underexposed or overexposed as to be undiagnostic.

Exposure Latitude

Exposure latitude refers to the range of exposure mAs values for a given x-ray beam penetration that will produce radiographic densities within the accepted diagnostic viewing range of .25 to 1.25 + base + fog. Latitude at its two extremes is categorized as being either wide or narrow. *Wide latitude* means that acceptable radiographs can be produced within a fairly broad range of exposure settings. *Narrow latitude* means that acceptable radiographs may be produced within a limited range of exposure settings. At a higher kVp the x-ray beam is even more heterogeneous. With lower kVp, the beam is less heterogeneous. At lower kVp settings, higher mAs values must be used to compensate for the loss in density. Two sets of films illustrate this point. The set of five skulls presented in Figure 5-8 were all taken at 85 kVp. Figure 5-9 contains five more skull films using a low kVp of 60. All were taken with a radiographic phantom. The middle film in each set is the "control" and is the one from which the films on either side are varied. Films to the left and right of center (B and D) represent -10% mAs and +10% mAs, respectively, while films A and E represent -20% mAs and +20% mAs, respectively.

Note that each set of films varies by the same percentage from the control. As expected, the films in Figure 5-8 exhibit the grayish look of low contrast; however, through the entire 40% range of mAs (20% each direction from the control), all films are satisfactory. Immediately obvious in Figure 5-9 is the classical "sharp" look of high contrast. Here we note, ironically, that only the middle three films (B-D) would probably be judged as acceptable.

The conclusion, therefore, is that with a marginally insufficient kVp, the exposure latitude is very narrow and there is little room for exposure error by the radiographer. With a higher kVp, the exposure latitude is wider and the radiographer has a greater range in the selection of appropriate mAs. As a result exposure errors decrease.

The Effects of kVp on Image Formation

KVp does not have an effect on the geometric properties of recorded detail, size distortion, or shape distortion of image

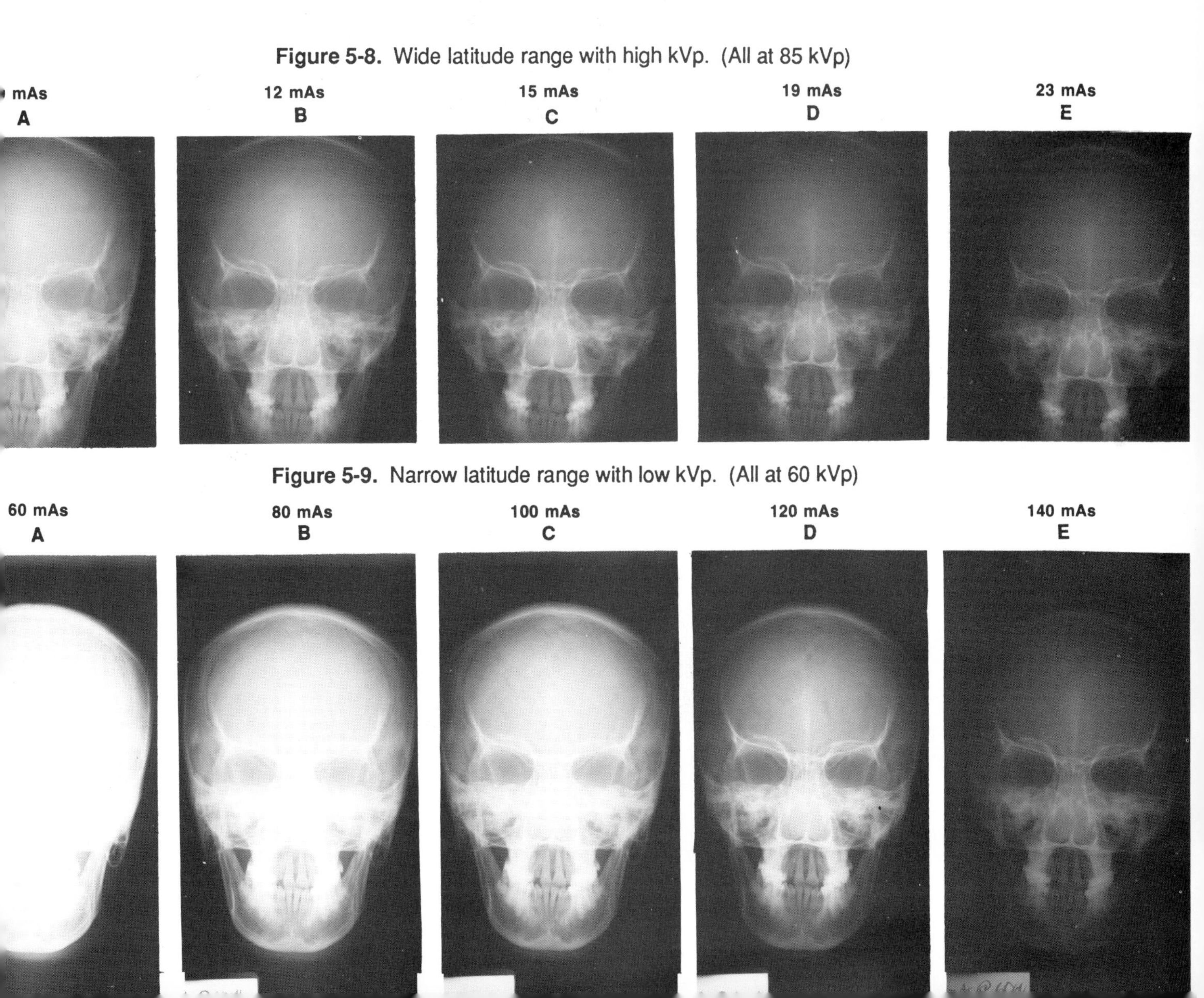

Figure 5-8. Wide latitude range with high kVp. (All at 85 kVp)

Figure 5-9. Narrow latitude range with low kVp. (All at 60 kVp)

details. Changing the kVp may make a difference in visibility, but not in the geometric properties of the film.

SUMMARY

This chapter has described the importance of kilovoltage to technique selection. Kilovoltage is the factor determining the force pushing and pulling electrons across the x-ray tube. A higher kilovoltage produces more x-rays and each will have more energy, a shorter wavelength, and greater penetrating power. A lower kVp will produce fewer x-rays and each will have less energy, a longer wavelength, and less penetrating power.

Kilovoltage affects radiographic density, contrast, and latitude. Kilovoltage has a direct effect on density. A higher kVp will produce a radiograph with more density. Lower kVp will produce a lighter radiograph.

Kilovoltage has an inverse effect on radiographic contrast. Higher kVp will produce less contrast. Since films exhibiting less contrast have more shades of gray, it would also be correct to say that kVp has a direct relationship to the scale of contrast. Scale of contrast refers to the number of shades of gray within the density spectrum of the image. A higher kVp will produce a radiograph exhibiting a longer scale of contrast (more shades of gray). Lower kVp will produce more (or higher) contrast. Lower kVp can also be said to produce a film with a shorter scale of contrast (fewer shades of gray within the spectrum).

Kilovoltage directly affects the exposure latitude. This means that at a higher kVp, there is a greater tolerance for mAs variations. At a high kVp the exposure latitude is wide. With low kVp the exposure latitude is narrow.

Kilovoltage does not affect recorded detail, size distortion, or shape distortion.

Table 5-1 has been provided as a review of the effects that kilovoltage has on radiographic quality.

TABLE 5-1.
Summary of the Effects of kVp on Radiographic Quality.

IMAGE VISIBILITY	
Radiographic Density	+
Radiographic Contrast	–
IMAGE FORM	
Recorded Detail	0
Size Distortion	0
Shape Distortion	0

+ = Direct Relationship to Image Characteristic
– = Inverse Relationship to Image Characteristic
0 = No Effect on Image Characteristic

STUDY QUESTIONS

1. What is the role of kilovoltage within the x-ray tube?
2. As kVp is changed, what will be the effect on the x-rays produced?
3. Explain the "rule of 10."
4. As kVp is changed from 60 to 70, what if anything, will happen to radiographic density?
5. As kVp is changed from 90 to 80, what if anything, will happen to subject contrast?
6. As kVp is changed from 90 to 80, what if anything, will happen to the scale of contrast?
7. As kVp is increased, what, if anything, will happen to exposure latitude?
8. As kVp is changed from 75 to 60, what if anything, will happen to image unsharpness?
9. As kVp is changed from 50 to 60, what if anything, will happen to image size distortion?
10. As kVp is increased (within appropriate limits) why does the scale of contrast increase? What will happen to radiographic density?

BIBLIOGRAPHY

1. Cahoon, John B. *Formulating X-Ray Techniques*, 8th ed., St. Louis: C. V. Mosby Company, 1980.
2. Donohue, Daniel P. *An Analysis of Radiographic Quality*. Baltimore: University Park Press, 1980.
3. Jenkins, David. *Radiographic Photography and Imaging Processes*. Baltimore: University Park Press, 1980.
4. *The Fundamentals of Radiography*. Rochester: Eastman Kodak Company, 1980.

Chapter 6

The Effects of Milliampere-Seconds on Radiographic Quality

Chapter Outline

1. Learning Objectives
2. Introduction
3. The Role of Milliampere-Seconds in Radiography
4. The Effects of Milliampere-Seconds on Image Visibility
   A. Radiographic Density
   B. Radiographic Contrast
5. The Effects of Milliampere-Seconds on Image Formation
   A. Recorded Detail
6. Summary
7. Study Questions

Learning Objectives

Upon completion of this chapter, the student should be able to:

1. Explain the role of milliamperage.
2. Explain the role of seconds.
3. Calculate mAs when given two known factors.
4. Determine if an exposure is safe for the x-ray tube by using a tube-rating chart.
5. Convert fractions to decimals by using a conversion chart.
6. Calculate the appropriate mAs for a change in density.
7. Explain how mAs affect radiographic density.

Introduction

The most challenging aspect of medical radiography is the control the radiographer has over the exposure technique factors. The radiographer must understand the function and relationship of the different exposure factors to the imaging characteristics of the radiograph. One does not have to be in the field of radiology very long to realize that more than one technical factor can significantly affect an imaging characteristic. The problem for the radiographer is to correctly select optimal combinations of factors which should be changed to obtain the optimum radiographic quality. One of these important factors is mAs.

The Role of Milliampere-Seconds in Radiography

If an exposure factor is changed, more than one radiographic imaging characteristic may be noticed. Selection of an exposure factor that will change only one descriptive characterstic at a time is usually best. Milliampere-seconds (mAs) is such a factor. The radiographer needs to know exactly what mAs is, how to calculate mAs, and how to determine if a correct mAs setting is achieved.

One must look at two separate exposure factors in order to fully understand milliampere-seconds (mAs). The first of these factors is milliamperage (mA). Milliamperage (mA) is equal to one-thousandth of an ampere. When radiographers select a given mA station, they are controlling the amount of current (electrons) that will flow through the negative cathode filament wire of the x-ray tube. Let's quickly review the different components of the x-ray tube to ensure an understanding of the part that milliamperage plays in the production of x-rays.

There are two ends of an x-ray tube: the negative or cathode and the positive or anode. Each end of the x-ray tube housing has different structures that are vital to the production of x-rays. For x-rays to be produced there are three basic requirements:

1. a source of electrons (filament/cathode)
2. acceleration of electrons (kVp)
3. stoppage of the electrons (target/anode)

The source of electrons comes from the filament wires located at the cathode end of the x-ray tube. To control the number of electrons being produced, one needs to control the current through the filament wires. Changes in the milliamperage (mA) will regulate the amount of current flowing to the filament. The radiographer has control of how much current will flow to the filament, thus controlling the number of electrons available to be made into x-rays.

Current flowing through the filament causes the filament to heat up and glow red. This heating of the filament is also known as thermionic emission (boiling off of electrons). The hotter the filament becomes, the greater the number of electrons "boiled off." The filament of the x-ray tube is similar in some ways to the filament of a toaster. When one places bread into the toaster, the bread is lowered and the current starts to flow in the toaster's filament wires. The filament glows red and there is a boiling off of electrons which causes the bread to heat, thus turning it brown in color. The radiographer can control the production of electrons by changing the mA located on the control panel. Controlling the number of electrons will determine the amount of x-rays in the primary beam. Increasing mA will increase electron production; decreasing mA will decrease electron production. There is, therefore, a direct relationship between mA and the production of electrons.

An increase of mA will add overall density or blackness to a radiograph, and vice versa. By increasing the mA, more electrons will be produced and thus more x-rays will be produced in the primary beam. Figure 6-1 shows two radiographs of a PA projection of the skull. Radiograph A used the technical factors of 85 kVp, 100 mA, 1/20s, 40″ FFD. One can readily see the radiograph is underexposed; it needs more density. In Radiograph

Figure 6-1.

Radiograph A
PA skull 85 kVp, 100 mA, 1/20 S, underexposed, needs more density

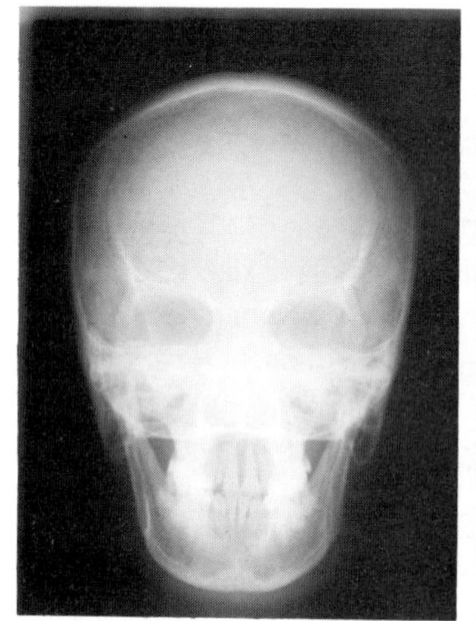

Radiograph B
PA skull 85 kVp, 200 mA, 1/20 S, exposure correct, has sufficient density.

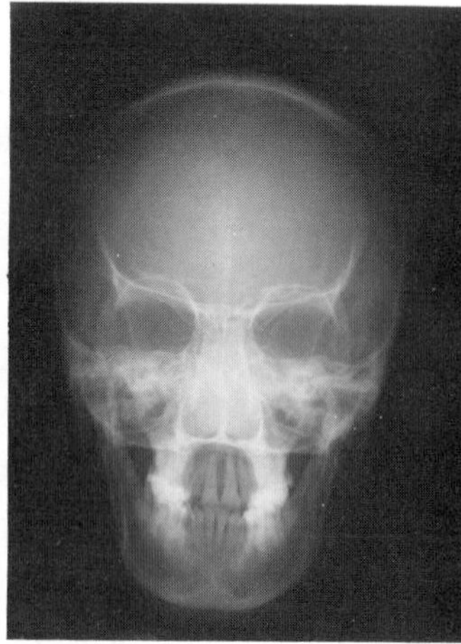

Figure 6-2.

Radiograph A
50 kVp, 100 mA, 1/40 S, 40" FFD, film needs more radiographic density.

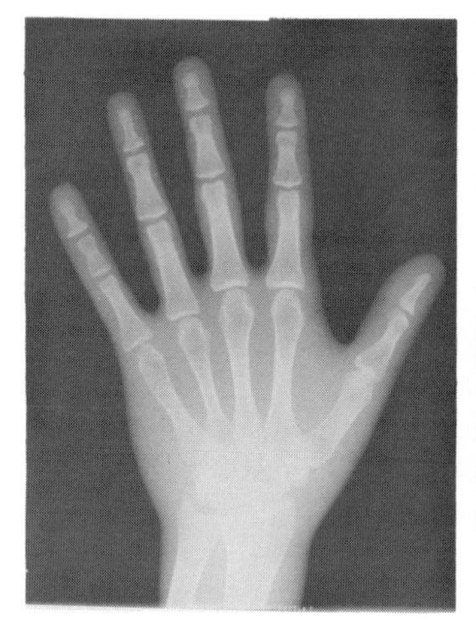

Radiograph B
50 kVp, 100 mA, 1/20 S, 40" FFD, Film has more radiographic density.

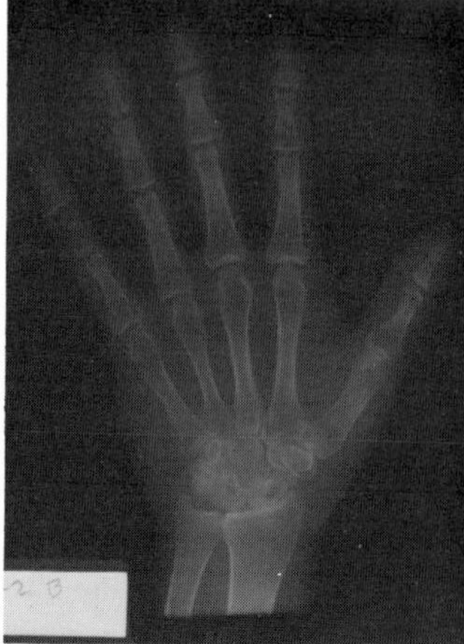

B, the mA was increased to 200 mA (that is, doubled) and all other technical factors remained the same. The amount of density on the film has been increased, thus producing a better radiograph.

The second factor of mAs is seconds or time. Exposure time controls the length of the x-ray exposure or how long the electrons will be allowed to travel across the x-ray tube from cathode to anode. Longer exposure times result in more electrons traveling across the tube. Thus, the quantity of x-ray in the primary beam will increase. One has to be careful when changing time. Increasing time also means the part being radiographed will have more of an opportunity to move during the exposure which will cause motion and a loss of recorded detail of the structural lines.

Figure 6-2 shows a posterior-anterior projection of the hand. Radiograph A used the exposure factors of 50 kVp, 100 mA, 1/40s, 40″ FFD. The hand in Radiograph A has been fully penetrated, but the radiograph is underexposed; it needs more density. Radiograph B used the exposure factors of 50 kVp, 100 mA, 1/20 s, 40″ FFD. The time was lengthened (i.e., doubled) from 1/40 to 1/20 of a second. One can see the amount of density on the radiograph has been increased.

Miliamperage multiplied by seconds equals milliampere seconds (mAs). These two factors go hand in hand and are impossible to separate. When asked for technical factors, most radiographers will usually state a certain amount of kVp and mAs rather than mA and seconds. Technique charts will often call for a certain amount of mAs, and the radiographer must then determine how much mA and time should be used.

Generally speaking, any combination of mA and S that will produce a certain amount of mAs will give the same amount of density on a radiograph. Figure 6-3 shows four radiographs of an anterior-posterior view of the skull. All of the radiographs used the same amount of kVp and mAs. The only factors changed were the mA and S, but all four combinations of mA and S equalled 25 mAs. One can see that the amount of density on all four films is identical.

It should be stressed that if the part being radiographed is not sufficiently penetrated, increasing the mAs will not produce a satisfactory film. Figure 6-4 shows four radiographs of a posterior-anterior projection of the skull. All of the radiographs use the same amount of kVp. The mAs, however, was varied. Radiograph A used 55 kVp, 50 mAs, 40″ FFD producing an underexposed radiograph. Radiograph B used 55 kVp, 100 mAs, 40″ FFD, and the radiograph is still underexposed. In Radiograph C and D the mAs was increased again and the kVp remained the same. The same results occurred. What the original radiograph needed was more penetration (kVp). If the anatomical part is not sufficiently penetrated, then increasing mAs will *not* result in a good radiograph.

Radiographers know that mA x s equals mAs, but how can one figure out the proper mA or s if only the mAs and one other factor is known? mA x s = mAs is an equation. If two parts of the equation are given, the ''unknown'' part can be found using basic algebra. The following problem is a situation that many radiographers face daily. A radiographer is going to perform an anterior-posterior projection of the distal femur which requires 70 kVp, 20 mAs, 40″ FFD. The x-ray machine has on the control panel 50 mA, 100 mA, and 200 mA stations. The radiographer decides to use a short exposure time because the patient is unable to cooperate. What is the shortest time that could be used with the existing mA stations? Since mAs is the product of mA and S, a relatively high mA would imply a relative short exposure time. The radiographer decides to use the highest mA station possible which is 200 mA. The following shows the proper mathematical computation for determining exposure time using the mAs formula:

$$mA \times s = mAs$$

$$\frac{200\ s}{200} = \frac{20}{200} = s = 1/10$$

The proper exposure time for this mA station, and the shortest exposure time possible would be 1/10 of a second. Radiographers have to make the final decision of the proper mA and time. They are the ones in control of the radiographic equipment and are aware of the circumstances that involve that particular patient and procedure.

Beginning radiographers may have a hard time conceptualizing the calculation of a proper mAs. The use of a memory wheel as in Figure 6-5 is an aid for working and using the mAs formula. Notice that mAs is on top with mA and s on the bottom. Covering up mAs leaves mA and s side by side. When there are two factors side by side, they are multiplied. The same is true on the wheel, mA x s = mAs. Covering up the mA (bottom left), that leaves mAs on top and s on the bottom. When one factor is on top of another these factors are divided into each other.

Figure 6-3.

Radiograph A
50 kVp, 25 mA,
1 s = 25 mAs

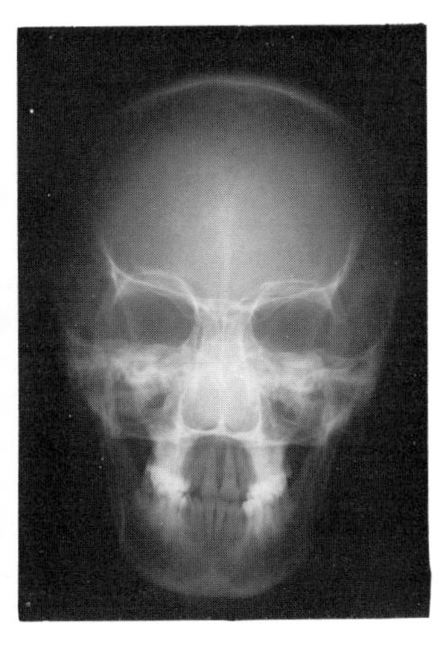

Radiograph B
50 kVp, 50 mA,
1/2 s = 25 mAs

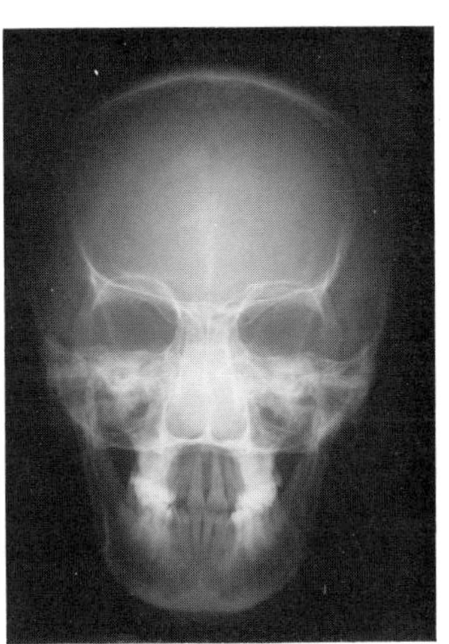

Radiograph C
50 kVp, 100 mA,
1/4 s = 25 mAs

Radiograph D
50 kVp, 200 mA,
1/8 s = 25 mAs

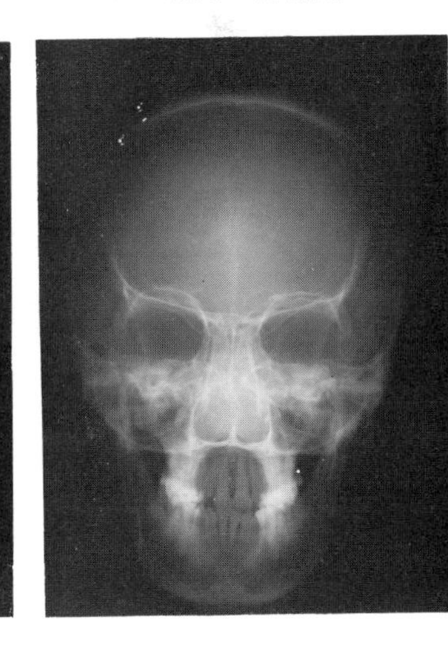

Figure 6-4.

Radiograph A
55 kVp, 50 mAs,
not enough density

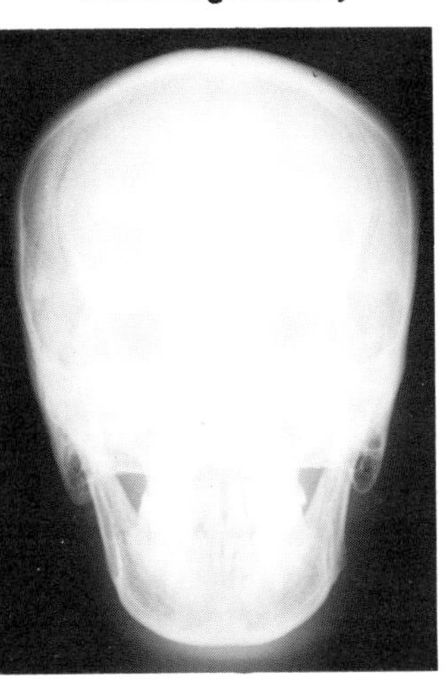

Radiograph B
55 kVp, 100 mAs,
still not enough density

Radiograph C
55 kVp, 150 mAs,
needs more density

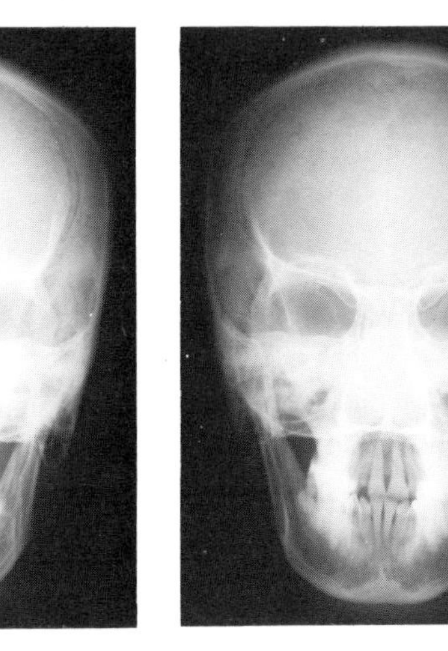

Radiograph D
55 kVp, 200 mAs,
needs more density

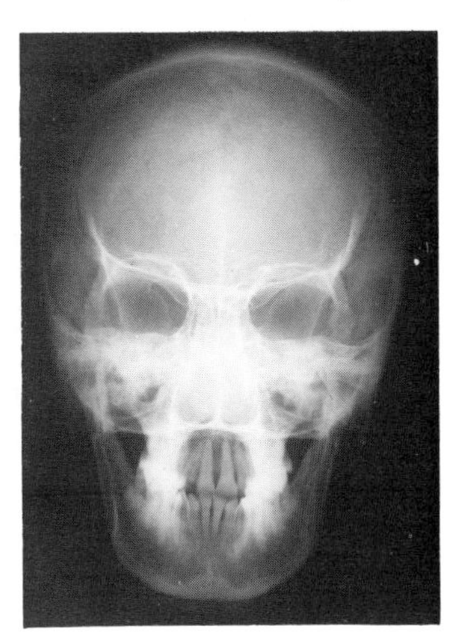

Figure 6-5. Milliampere Seconds (mAs) Memory Wheel. (Cover missing variable and work algebraic operation.)

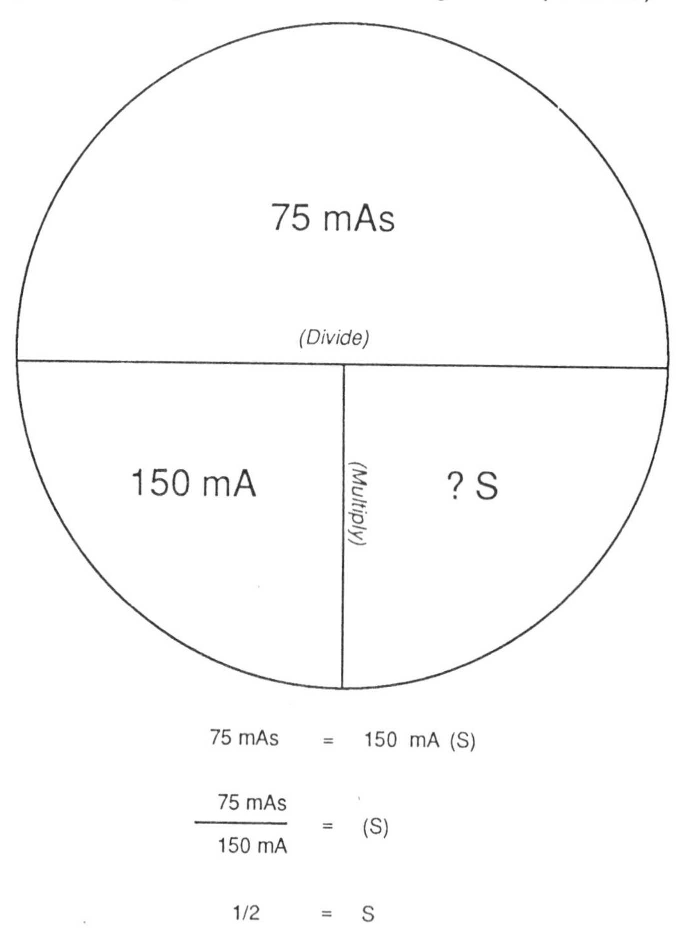

The same is true on the memory wheel, and mAs would be divided by s in order to find mA.

To use the memory wheel the following steps should be taken:

1. The wheel must be properly set up.
2. Insert the known information.
3. Cover up the unknown information (the factor that needs solving); and
4. Look at the two remaining factors. If they are side by side, then multiply. If one factor is on top of the other, then divide the lower one into the top number.

If a radiographer is filming an anterior-posterior projection of the lumbar spine that requires 75 mAs and decides to use 150 mA, what time should be used to produce 75 mAs? Following the above instructions, one can see that 75 mAs needs to be divided by 150 mA or it could be put into a fraction of 75/150. Either way it will work out to be 1/2 (.5) meaning the time on the machine should be set on 1/2 (.5) second. To prove that 1/2 second is correct, multiply 150 mA x 1/2 S = 75 mAs.

Another tool used to identify mAs by various mA and time combinations is the use of a mAs chart. Figure 6-6 represents such a chart. The first two columns on the chart represent time—one in impulses the other in fractions. The rest of the columns represent different mAs values, and the last column at the bottom of the chart represents mA. To use the chart select the desire mAs relative to a given mA vlaue and then read to the left sid of the chart to determine the time. If one needs a short exposur time, for example 5 mAs, the shortest exposure is 1/60 of a sec ond. Read across the 1/60 line until 5 mAs is found, then rea downward to the bottom of the chart until it intersects with th mA line, which is 300 mA. If the x-ray machine does not hav a 300 mA station, then select an available mA station. For ex ample, try the 100 mA. Read up the mA column until 5 mA is found then read to the left side of the chart until it intersect with the time, which is 1/20 of a second or 6 impulses. This cha is easily adaptable to any x-ray equipment.

Reviewing the mAs conversion chart results in another apparer problem; the time is in impulses or fractions. But what about th x-ray machine time that is in decimals? Figure 6-7 represents chart for converting fractions into decimals. The horizonta numbers of the chart (1-12) represent the numerator of the frac tion. The vertical numbers represent the denominators. To rea the chart for converting the fraction 3/8 to a decimal, first tak the numerator (3) and read down on the column labeled 3 unti it intersects with the denominator (B) line. Then read the numbe at that intersection which is the decimal equivalent—.375. A con venience for the radiographer, mAs charts are available wit decimal exposure times eliminating the extra step of convertin a fraction to a decimal. These different charts are all tools to mak the radiographer's job a littler easier. Remember, charts are on ly as good as the individual using them.

When mA or s is used to compensate for abnormal situation in the clinic, radiographers need to be aware of what is happen ing in the production of electrons and x-rays. Increasing the mA increases the current flowing to the filament producing more elec trons. A minimum kV level must be used not only to penetrate the body part but also to efficiently use all the available elec trons established by the mA value selected. If a minimum kV is not used, the electrons being emitted repel, or hold back, the electrons just about to be emitted. This is called a space charge effect. This problem is associated with higher mA settings anc the radiographer should be aware that when using higher mA settings the proper kV must also be used.

Increasing the mA puts a greater limitation on the x-ray tube since one is limited in the seconds and kVp that can be used with a given exposure mAs. This is why tube-rating charts were designed and there is a chart specific to every x-ray tube in a radiology room and for each focal spot of the tube (s). Figure 6-8 illustrates a tube-rating chart that shows the different kV, s, and mA that can be safely used without damaging the x-ray tube. Here is how the tube-rating chart is read: At the intersection of a given kV and s, place a point. Look for the given mA line. If the point is to the left of the mA line, it is a safe exposure. If the point is to the right of the mA line, it is an unsafe exposure.

Using the tube-rating chart in Figure 6-8 and the following technique of 85 kVp, 200 mA, 1/10 s, determine if the exposure is safe for the x-ray tube. Placing a point at the intersection of 85 kVp, and 1/10 s, look for the 200 mA line. The point is to the left of the 200 mA line so it is a safe exposure. Notice that

Figure 6-6. mAs Conversion Chart.

Impulses	Time (seconds)	Milliampere seconds											
2	1/60	0.16	0.25	0.33	0.41	0.50	0.83	1.66	2.50	3.33	5.00	6.66	8.33
3	1/40	.25	.37	.50	.62	.75	1.25	2.50	3.75	5.00	7.50	10.00	12.50
4	1/30	.33	.5	.66	.83	1.00	1.66	3.33	5.00	6.66	10.00	13.33	16.66
5	1/24	.41	.62	.83	1.04	1.25	2.08	4.12	6.25	8.33	12.50	16.66	20.83
6	1/20	.50	.75	1.00	1.25	1.50	2.50	5.00	7.5	10.00	15.00	20.00	25.00
7		.58	.87	1.16	1.56	1.75	2.91	5.83	8.75	11.66	17.50	23.33	29.16
8	1/15	.66	1.00	1.33	1.66	2.00	3.33	6.66	10.00	13.33	20.00	26.66	33.33
9	3/40	.75	1.12	1.50	1.87	2.25	3.75	7.50	11.25	15.00	22.50	30.00	37.50
10	1/12	.83	1.25	1.66	2.08	2.50	4.16	8.33	12.5	16.66	25.00	33.33	41.66
11		.91	1.37	1.83	2.27	2.75	4.58	9.16	13.75	18.33	27.50	36.66	45.83
12	1/10	1.00	1.5	2.00	2.50	3.00	5.00	10.00	15.00	20.00	30.00	40.00	50.00
13		1.08	1.62	2.16	2.77	3.25	5.41	10.83	16.25	21.66	32.50	43.33	54.16
14		1.16	1.75	2.33	2.91	3.50	5.83	11.66	17.50	23.33	35.00	46.66	58.33
15		1.25	1.87	2.50	3.12	3.75	6.25	12.50	18.75	25.00	37.50	50.00	62.50
16	2/15	1.33	2.00	2.66	3.33	4.00	6.66	13.33	20.00	26.66	40.00	53.33	66.66
17		1.41	2.12	2.83	3.54	4.25	7.08	14.16	21.25	28.33	42.50	56.66	70.83
18	3/20	1.50	2.25	3.00	3.74	4.50	7.50	15.00	22.50	30.00	45.00	60.00	75.00
19		1.58	2.37	3.16	3.95	4.75	7.91	15.83	23.75	31.66	47.50	63.33	79.16
20		1.66	2.50	3.33	4.16	5.00	8.33	16.66	25.00	33.33	50.00	66.66	83.33
21		1.75	2.62	3.50	4.37	5.25	8.75	17.50	26.25	35.00	52.50	70.00	87.5
22		1.83	2.75	3.66	4.58	5.50	9.16	18.33	27.50	36.66	55.00	73.33	91.66
23		1.91	2.87	3.83	4.79	5.75	9.58	19.16	28.75	38.33	57.50	76.66	95.83
24	1/5	2.00	3.00	4.00	5.00	6.00	10.00	20.00	30.00	40.00	60.00	80.00	100.00
25		2.08	3.12	4.16	5.20	6.25	10.41	20.83	31.25	41.66	62.50	83.33	104.16
26		2.16	3.25	4.33	5.41	6.50	10.83	21.66	32.50	43.33	65.00	86.66	108.33
27		2.25	3.37	4.50	5.62	6.75	11.25	22.50	33.75	45.00	67.50	90.00	112.50
28		2.33	3.50	4.66	5.83	7.00	11.66	23.33	35.00	46.66	70.00	93.33	116.66
29		2.41	3.62	4.83	6.04	7.25	12.08	24.16	36.25	48.33	72.50	96.66	120.83
30	1/4	2.50	3.75	5.00	6.25	7.50	12.50	25.00	37.50	50.00	75.00	100.00	125.00
Milliamperes		10	15	20	25	30	50	100	150	200	300	400	500

as mA is increased, the amount of kV that can be safely used is decreased. Also, as the mA is increased, the number of time choices is decreased. It is true that mA x s = mAs, but the maximum mAs one can safely produce using the 400 mA station is approximately 80 mAs. Using the 200 mA station, the maximum mAs that can be safely produced is approximately 800 mAs.

Milliamperage is not always the best way to change the amount of density on a radiograph. What would happen if time were changed? If you recall, changing time will control how long the electrons will flow across the x-ray tube. The largest problem with increasing time to increase density is that the part being radiographed would have more opportunity to move thus producing motion on the radiograph. If the patient understands the instructions that are given, and is physically able to cooperate, this is not usually a problem. One of the advantages of changing time is the wide selection of options. Most x-ray equipment will have a greater variety of time compared to the selection of mA stations. Sometimes only a slight increase or decrease is needed and this is easier to obtain by changing time instead of mA. When using single-phase equipment, a change of the mA could mean changing from a small focal to a large focal spot with a loss of some recorded detail.

The Effect of Milliampere-Seconds on Image Visibility

Radiographic Density

Density is directly proportional (linearly) to milliampere-seconds (mAs). This means that if mAs is changed by 10 percent, the amount of density on a radiograph will also be changed by 10 percent. This presents a difficult problem for radiographers. How much should one change the mAs to really change the density on a radiograph? Studies that have been conducted report that a change of 20 to 35 percent in mAs must be made to achieve a *visible* change in density.

Figure 6-7. Decimal Equivalents of Fractional Exposures.

Denominators	Numerators: 1	2	3	4	5	6	7	8	9	10	11	12	Denominators
2	0.5	1.	1.5	2.	2.5	3.	3.5	4.	4.5	5.	5.5	6.	2
3	.333	.666	1.	1.33	1.666	2.	2.33	2.66	3.	3.33	3.66	4.	3
4	.25	.5	.75	1.	1.25	1.5	1.75	2.	2.25	2.5	2.75	3.	4
5	.2	.4	.6	.8	1.	1.2	1.4	1.6	1.8	2.	2.2	2.4	5
6	.167	.333	.5	.667	.835	1.	1.167	1.333	1.5	1.667	1.833	2.	6
7	.143	.286	.429	.572	.715	.858	1.	1.43	1.286	1.429	1.57	1.7	7
8	.125	.25	.375	.5	.625	.75	.875	1.	1.125	1.25	1.375	1.5	8
9	.111	.222	.333	.444	.555	.666	.777	.888	1.	1.111	1.222	1.333	9
10	.1	.2	.3	.4	.5	.6	.7	.8	.9	1.	1.1	1.2	10
11	.09	.18	.27	.363	.455	.545	.636	.727	.818	.090	1.	1.09	11
12	.083	.167	.25	.333	.415	.5	.583	.667	.75	.833	.917	1.	12
15	.067	.134	.2	.267	.333	.4	.467	.533	.6	.667	.733	.8	15
20	.05	.1	.15	.2	.25	.3	.35	.4	.45	.5	.55	.6	20
24	.042	.083	.125	.167	.208	.25	.292	.333	.375	.416	.458	.5	24
30	.033	.067	.1	.133	.167	.2	.233	.267	.3	.333	.367	.4	30
40	.025	.05	.075	.1	.125	.15	.175	.2	.225	.25	.275	.3	40
60	.017	.033	.05	.067	.083	.1	.117	.133	.15	.167	.183	.2	60
120	.008	.017	.025	.033	.042	.05	.058	.067	.075	.083	.092	.1	120

One may ask why there is such a discrepancy in the percentages? Most studies will use words like a "noticeable," "significant," and "considerable" change of density when changing mAs by a certain percent. But, what do these words mean? They are really "abstract" or "opinionated" or "subjective" in nature; placing one's own value to them. Through experience, radiographers are able to make acceptable adjustments in radiographic density using mAs deviations from the norms established by departmental technique charts.

As stated, changing mAs will change the number of electrons in a given exposure. This alteration of mAs will affect the number of x-rays in the primary beam as well as the amount of x-rays to the film, causing a change in the density. Figure 6-9 presents four radiographs of an anterior-posterior projection of the skull. Each radiograph had an increase of mAs. For this study a change of 20 percent in mAs was used. Radiograph A used the technical factors of 75 kVp, 13 mAs, 40" FFD and produced an acceptable radiograph. If the radiologist wanted to see more density on the radiograph, how much more mAs should be used? The radiographer decided to increase the original mAs by a factor of 20 percent and used the technical factors of 75 kVp, 16 mAs, at 40" FFD. The results of this change are demonstrated in Radiograph B. The density has increased slightly as one can see a noticeable change. The same concept is demonstrated in Radiographs C and D. The mAs was increased 20 percent in each radiograph. Radiograph C used the factors of 75 kVp, 20 mAs, 40" FFD and Radiograph D used the factors of 75 kVp, 26 mAs, 40" FFD. In each radiograph there is a noticeable increase in density.

Figure 6-8. Tube Rating Chart.

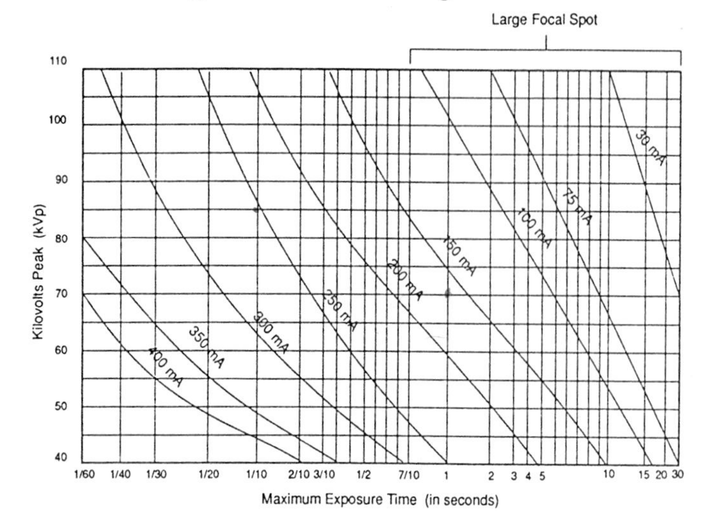

Density is directly proportional to the amount of mAs. Doubling the amount of density on a radiograph requires doubling the amount of the original mAs. Figure 6-10 shows two radiographs of a posterior-anterior view of the hand. Radiograph A used the technical factors of 48 kVp, 1.2 mAs at 40" FFD. Radiograph B used the technical factors of 48 kVp, 2.4 mAs, at 40" FFD. Radiograph B compared to Radiograph A demonstrates twice as much radiographic density.

Figure 6-9.

Radiograph A
75 kVp, 13 mAs,
40" FFD.

Radiograph B
75 kVp, 16 mAs
40" FFD, slight increase
in density

Radiograph C
75 kVp, 20 mAs,
40" FFD, slight increase
in density

Radiograph D
75 kVp, 25 mAs,
40" FFD, slight increase
in density

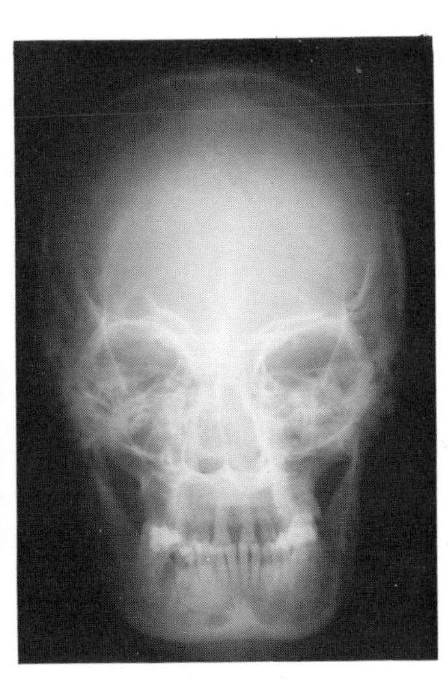

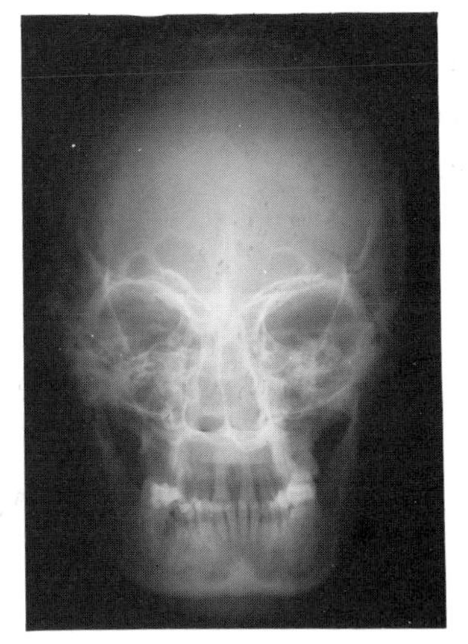

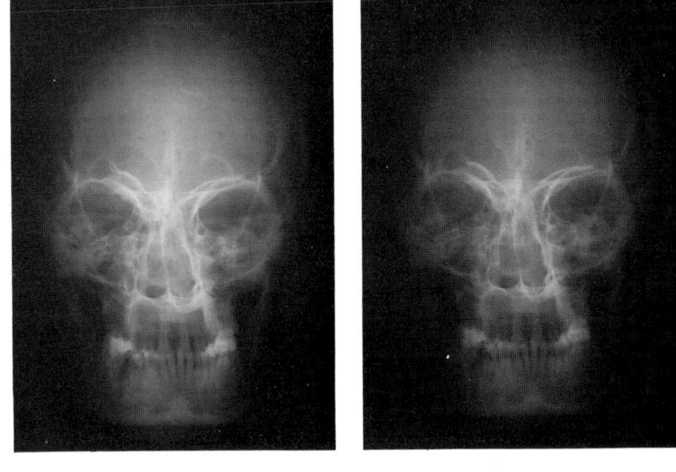

Figure 6-10.

Radiograph A
48 kVp, 1.2 mAs,
40" FFD.

Radiograph B
48 kVp, 2.4 mAs,
40" FFD. Amount of
density is double.

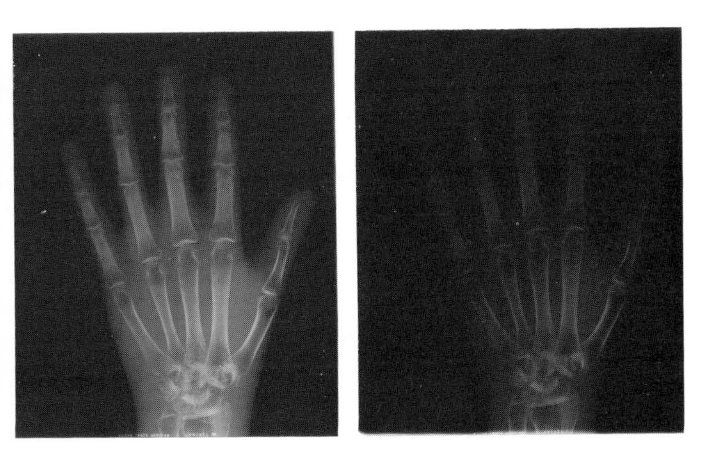

Radiographic Contrast

Radiographic contrast has already been defined as the measureable or observed difference between adjacent radiographic densities and was discussed in Chapters 2 and 5.

When optimum kVp is used, a change of mAs will not usually affect the contrast. Reviewing Figure 6-9, notice how the change of mAs affects the density only. Comparing the four radiographs taken at different mAs values, the contrast remains the same. True, each radiograph has more density, but the various x-ray intensities bear the same relation to each other. For example, compare the glabella to the mandible. The difference between the shades of gray did not change, only the overall density.

The Effects of Milliampere-Seconds on Image Formation

Recorded Detail

Milliampere seconds can indirectly affect recorded detail. Remember, if mAs is used to increase the amount of density on a radiograph and time (seconds) is increased, the part being radiographed has more opportunity to move during the exposure. If the part moves, motion will be produced on the radiograph causing a blurring of the structure lines and a loss of recorded detail.

Another consideration on recorded detail is the Bloom Effect of the focal spot under high mA conditions. The ''blooming effect'' is caused by using a low kV and a high mAs. The focal spot increases directly to the tube current. The electrons will form a large cloud around the focal spot, thus increasing its size and resultant penumbral effect (blurring). Increasing the kV will help decrease the size of the cloud, decreasing the focal spot size.

Milliampere-seconds does not have an effect on the geometric properties of size distortion (magnification) or shape distortion (true).

SUMMARY

The following table (Table 6-1) shows the different interactions that milliampere-seconds has on the radiographic imaging process. Milliampere-seconds will affect the radiographic quality of the image details. An increase of the mAs will increase the amount of radiographic density of the radiograph; decrease the mAs and the amount of radiographic density will decrease. This is a direct relationship between mAs and radiographic density.

Milliampere-seconds has an indirect affect on unsharpness. If mAs must be changed and seconds are increased, the part being radiographed has more opportunity to move during the exposure. Movement during the exposure will cause the radiograph to be imaged with a fuzziness in the structural lines. Also consider Bloom Effect of the focal spot under high mA conditions and the effect on blurring.

Milliampere-seconds does not affect size distortion/magnification or shape distortion of the radiographic quality of the image details. When using optimum kVp techniques, a change of mAs will not affect the radiographic contrast.

TABLE 6-1. **Summary of the Effects of Milliampere-Seconds on Radiographic Quality.**

IMAGE VISIBILITY	
Radiographic Density	+
Radiographic Contrast	0
IMAGE FORM	
Recorded Detail	–
Size Distortion	0
Shape Distortion	0

+ = Direct Relationship to Image Characteristic
– = Inverse Relationship to Image Characteristic
0 = No Effect on Image Characteristic

STUDY QUESTIONS

1. Explain the function of mA in the radiographic process.

2. Explain how changes of time will affect the radiographic process.

3. Calculate three combinations of "mA" and "S" to produce 10 mAs. Your machine has the following choices for mA only: 50 mA 100 mA 150 mA 200 mA 300 mA 400 mA

4. Using the tube-rating chart in this chapter, determine if the following exposure is safe: 300 mA 75 kVp, 1/20 S.

5. Using the conversion chart in this chapter convert the fraction 5/8 into a decimal.

6. A radiograph required the exposure factors of 80 kVp, 15 mAs, 40″ FFD, small focal spot. When developed the radiograph revealed that the part was sufficiently penetrated but needed to have twice the amount of density. Compensate for the underexposed radiograph and calculate the new exposure factors.

7. Explain how mAs affects radiographic density, contrast, recorded detail, and size and shape distortion.

Chapter 7

The Effects of Distance on Radiographic Quality

Chapter Outline

1. Learning Objectives
2. Introduction
3. The Role of Distance in Radiography
   A. Inverse Square Law
3. The Effect of Distance on Image Visibility
   A. Radiographic Density
   B. Air Gap Technique
   C. Radiographic Contrast
4. The Effect of Distance on Image Formation
   A. Recorded Detail
   B. Size and Shape Distortion
5. Summary
6. Study Questions

Learning Objectives

Upon completion of this chapter, the student should be able to:

1. Define focal film distance and object-to-film distance.
2. Calculate the intensity of radiation at the new distance, given a change in distance.
3. Calculate the new mAs required to maintain radiographic density given a change in focal film distance.
4. Identify the effect, if any, of a change in focal film distance on any of the following:
   a. radiographic density
   b. radiographic contrast
   c. recorded detail
   d. size distortion
   e. shape distortion
5. Identify the effect, if any, of a change in object-to-film distance on any of the following:
   a. radiographic density
   b. radiographic contrast
   c. recorded detail
   d. size distortion
   e. shape distortion
6. Identify the FFD that would produce a radiograph with more (or less) density, more (or less) image sharpness, larger (or smaller) image size, as specified, given different focal film distance choices.
7. Identify the OFD that would produce a radiograph with more (or less) density, more (or less) image sharpness, larger (or smaller) image size, as specified, given different object-to-film distance choices.

Introduction

Distance plays a significant role in the making of diagnostic quality radiographs. Without standardized distance settings (in relationship to the quantity and quality of x-rays produced from a given x-ray tube) the resultant radiograph may be rendered too light, too dark, or blurred. Obviously, this can disrupt the diagnostic effort of the attending physician. Therefore, all medical radiographers must become aware of the distance factor in establishing optimum radiographic exposure parameters if they expect to produce diagnostic quality radiographic images.

The Role of Distance in Radiography

In radiographic discussions, terminology should be standardized so that everyone is working from the same base of understanding. The distance from the focal spot to the film is named differently by various sources. Some refer to that distance as TFD (target-film distance), (SID) source-image receptor distance, or anode-film distance, while others have long referred to it as the FFD (focal spot - film distance). For the purpose of this chapter the term focal film distance (FFD) will be used.

Focal film distance (FFD) can be illustrated as the distance from the flashlight bulb to the wall at which it is aimed. As FFD increases, the intensity (or brightness) decreases while the area covered increases proportionally to the square of the distances from the source. If a flashlight 1 foot from a wall is moved to 2 feet, there will be four times as much area covered by the beam. Since the same number of light photons are being emitted from the bulb covering this now larger area, the beam intensity at this longer distance is now only one fourth as great. This is actually a problem for radiographers since we would rather not have the film's exposure changed just because we had to use a different FFD.

Another factor that should be considered as an influence on radiographic quality is the distance from the patient to the film or image receptor. Again, there are different names to refer to this distance, such as object-to-film distance (OFD) and subject-receptor distance. For purposes of this chapter, object-to-film distance (OFD) will be used.

The effect of OFD can also be simply demonstrated. Using a flashlight held at a certain unchanging distance from a wall, place an object (such as a hand) between the flashlight and the wall. Pulling the hand away from the wall will cause the shadow (corresponding to the radiographic image) to increase in size, while at the same time the clarity of the silhouette will diminish. This compares of course, to the distance between the anatomical part being radiographed and the film or image receptor. There are many situations in which the body part is elevated or obliqued in such a manner that either the entire part or just a portion of it is not in direct contact with the film holder. To minimize magnification and also to provide for optimal structural definition, OFD is usually kept to a minimum.

Inverse Square Law

The change in light intensity caused by a distance change is the scientific principle known as the *inverse square law* and is one that most radiographers learn early.

One formula that will indicate this change in beam intensity is:

$$\frac{I}{i} = \left(\frac{d}{D}\right)^2$$

In this formula, I represents the new (unknown) intensity at a new distance of D^2; i represents the original known intensity at the original distance d^2. A problem requiring the use of this formula proposes that there is an exposure level of 14 milliroentgens (mR) at a 30-inch distance. What will be the intensity at a 40-inch distance? The solution to this is as follows:

$$\frac{I}{i} = \left(\frac{d}{D}\right)^2$$

$$\frac{I}{14} = \left(\frac{30}{40}\right)^2$$

$$\frac{I}{14} = \left(\frac{3}{4}\right)^2$$

$$\frac{I}{14} = (.75)^2$$

$$\frac{I}{14} = .5625$$

$$I = 14 \times .5625$$

$$I = 7.875 \text{ mR}$$

The above formula indicates the change in radiation intensity if distance changes when *no* adjustments in other factors are made. The responsibility of a radiographer is to make any changes necessary to balance radiographic quality even when situations like different FFD's are encountered (Figure 7-1). Portables, trauma cases, and others may require modified distances to be used between tube and film. Since the end result, the radiograph, is still to have the correct amount of exposure regardless of the distance change, tube output will have to be modified to keep the radiation level or beam intensity reaching the film at the original level.

Figure 7-1. Radiographic situation showing how radiographic density is affected by changes in target-to-film distance, according to inverse square law.

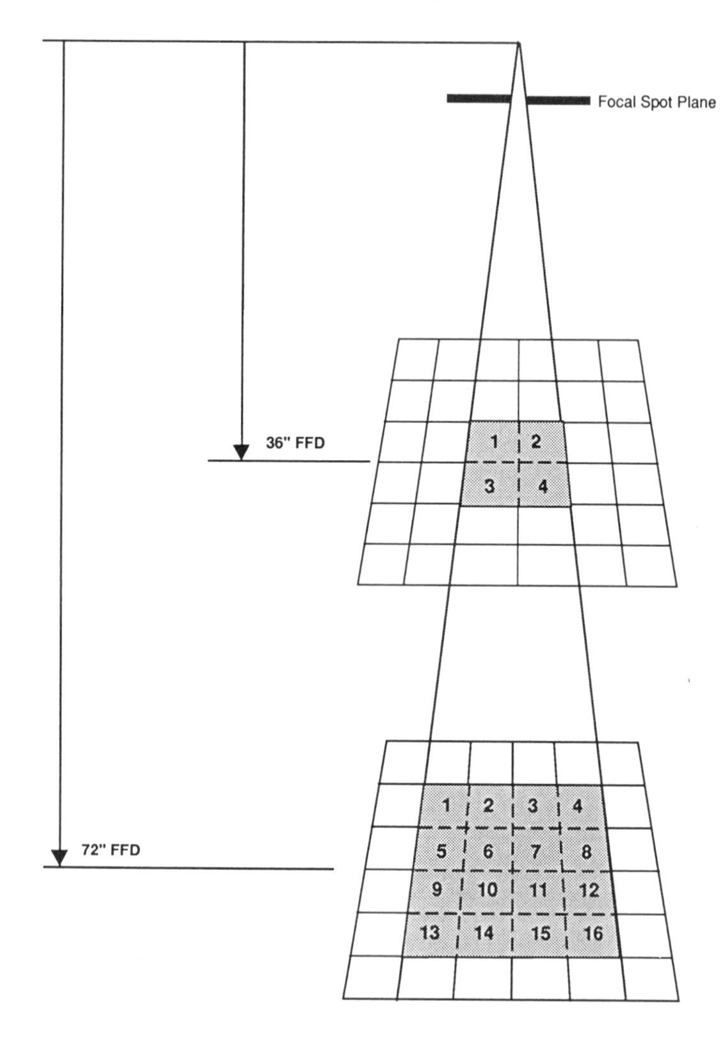

When intensity is altered the factor really being changed is radiation quantity. Other adjustment factors should address this kind of change rather than any modification in x-ray penetration. Adjustments are therefore made when possible in mAs rather than kVp.

The formula that radiographers must understand to correctly modify factors when FFD is changed is called the mAs-Distance Formula:

$$\frac{MAS_1}{MAS_2} = \left(\frac{FFD_1}{FFD_2}\right)^2$$

The subscripts beside the different variables are used as idenfiers where "1" represents the old factors and "2" represents he new changed element. In this formula, mAs is almost always he unknown variable and the one for which the formula must e solved. Notice also in this formula that the distances must be quared. Failure to recognize this is one of the most common nistakes in the solution of this formula. An example: if a situaion requires 50 mAs in a 40" FFD, how much mAs should be sed when the FFD is limited to 35"?

Using the formula stated above, the solution to this problem vould be as follows:

$$\frac{mAs_1}{mAs_2} = \left(\frac{FFD_1}{FFD_2}\right)^2$$

$$\frac{50}{mAs_2} = \left(\frac{40}{35}\right)^2$$

$$\frac{50}{mAs_2} = \left(1.1429\right)^2$$

$$\frac{50}{mAs_2} = 1.306$$

$$50 = mAs_2 \times 1.306$$

$$\frac{50}{1.306} = mAs_2$$

$$38.3 = mAs_2$$

One clue that you should have as to the relative amount of the ew mAs is the direction of the change in the FFD's. If distance s being reduced (since without other changes this would increase film density), mAs will have to be reduced to keep film density the same on the film. Since in the beginning mAs was 50 and use of the formula indicates that the new mAs will be about 38, a logical answer is realized. An answer of larger than 50 mAs would lead one to suspect that something as simple as a reverse assignment of FFD's had been made in solving the formula.

The Effect of Distance on Image Visibility

Radiographic Density

Focal film distance has an inverse effect on density. As FFD is increased, film density decreases (because of the inverse square law). As FFD decreases, film density increases. Note the set of hand radiographs in Figure 7-2, where FFD is the only variable changed. Since this is an unsolicited density change, density must usually be corrected by an mAs change to keep the overall density of the radiograph at its original level.

Object-to-film distance also affects density. The density of a radiograph is composed of primary and scatter radiation. If the amount of scatter radiation were to be reduced, the radiographic density would be less (Figure 7-3). This effect is less noticeable with thinner body parts.

Air-Gap Techniques

Chapter 13 will discuss grids as being one of the best methods of reducing scatter radiation. Through the use of an increased OFD, similar results can be obtained. Such a method is appropriately titled, the "air-gap technique." Air-gap technique consists of three steps. First, space is introduced between the film and patient. This is the "air-gap" and is most easily accomplished on upright views, such as chest radiography, usually on large patients. Second, FFD is increased to offset the detail and magnification problems that an increased OFD would usually cause. Third, mAs is recomputed as required by the mAs-distance formula.

Figure 7-2. Density as affected by target-to-film distance.
(Note that with shorter TFD, radiographic density increases; longer TFD results in less radiographic density.)

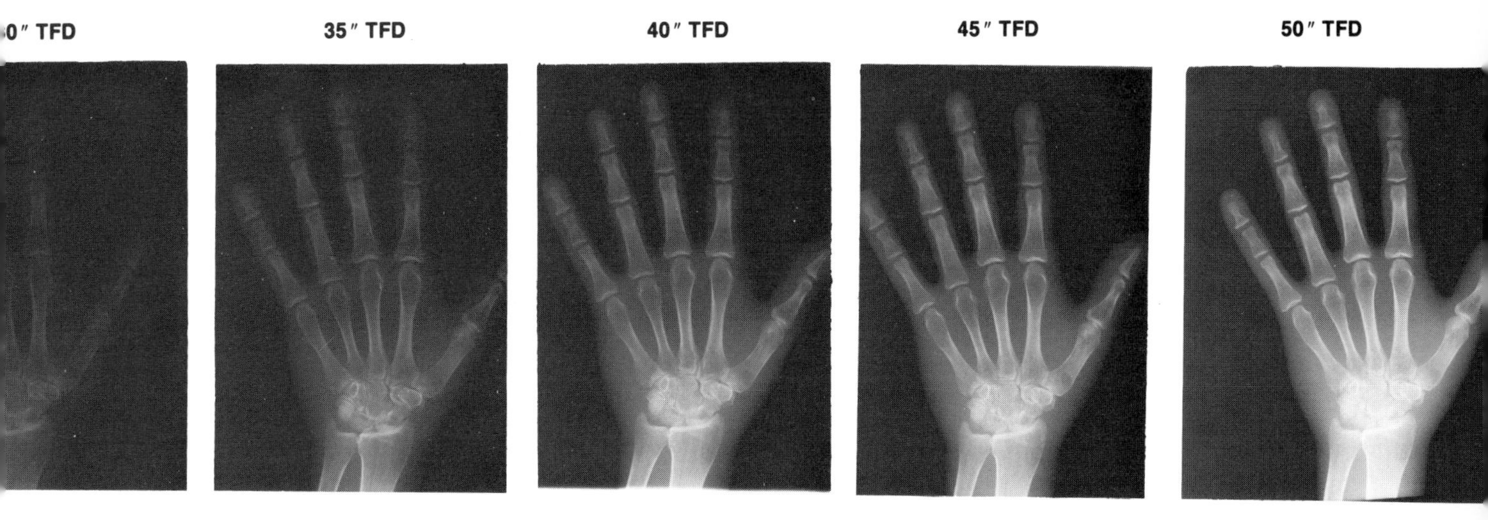

Figure 7-3. Object-to-film distance changes, because of the divergence of scatter x-rays, affect the amount of scatter reaching the radiograph.

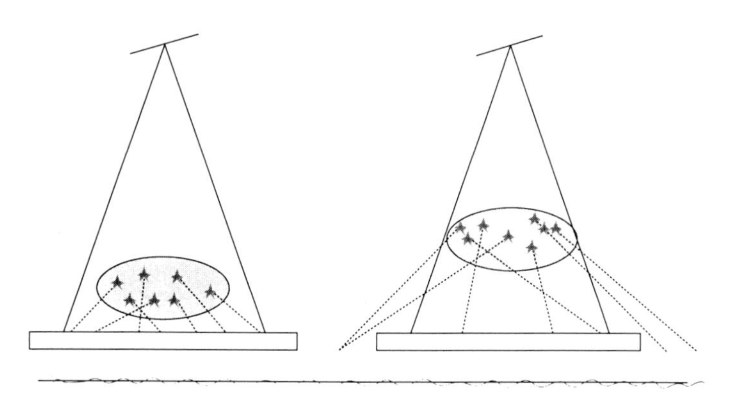

Air-Gap technique studies indicate that a 5-inch air gap is nearly as efficient as a 7:1 grid using a 10-centimeter thickness of water to represent a thin patient (Gould and Hale, 1974). With a 10-inch air gap, the method is as good as a 15:1 grid (this technique is commonly used in magnification radiography), but as patient thickness increases from 10 to 20 use of an air gap is not as efficient as the previously mentioned grids.

The relationship between both distances (OFD and FFD) and denstiy is inverse. More of either distance results in a decrease in density. Less of either distance results in an increase in density.

Radiographic Contrast

Focal film distance does not have an effect on radiographic contrast, but object-to-film distance definitely does. As OFD increases, the amount of scatter radiation reaching the film decreases, meaning that the radiograph will be devoid of much of the "fog" or grayness that it would otherwise possess (Figure 7-3). The radiograph, therefore, possesses greater image visibility, or more of a black vs. white look (higher contrast). OFD has a direct relationship to radiographic contrast, but like density is more noticeable with thicker body parts.

The Effect of Distance on Image Formation

Recorded Detail

Both distances have an effect on recorded detail. A simple experiment projecting the shadow of a hand onto a wall with a flashlight will quickly prove that the sharpness of the silhouette is improved with a longer distance between flashlight and wall. The projection of a body part, or anatomical structure, or a point on that anatomical structure would respond in the same way to changes in FFD (Figure 7-4). This is one reason that, as improvements in tube design and efficiency have been made, longer FFD's have become possible. This has resulted in an improvement in image definition or recorded detail. FFD, therefore, has an inverse relationship to recorded detail. A longer FFD results in more (or better) detail. A shorter FFD results in less recorded detail.

Object-to-film distance has an even more drastic effect on recorded detail. In the previously discussed experiment involving the hand and flashlight, try moving the hand closer and then further from the wall. (NOTE: Enlargement of the image is usually the first noticed change, but this is the distortion aspect of distance to be covered later.) Sharpness or clarity of the image silhouette, which can be compared to the detail of the radiographic image, will improve with a shorter OFD (Figure 7-5). With a longer OFD, there is less detail. Object-to-film distance, therefore, has an inverse relationship to recorded detail. (More OFD, less detail.) Figure 7-6 shows a radiographic example of this effect. When studying oblique projections of chest and rib views in the future, notice that one side of the patient's anatomy

Figure 7-4. Target-to-film distance changes as they affect image sharpness. Shaded area represents edges of the anatomical structure (sometimes referred to as penumbra); ideally, as small as possible. Rectangular area in the tube represents actual focal spot from which x-rays are produced.

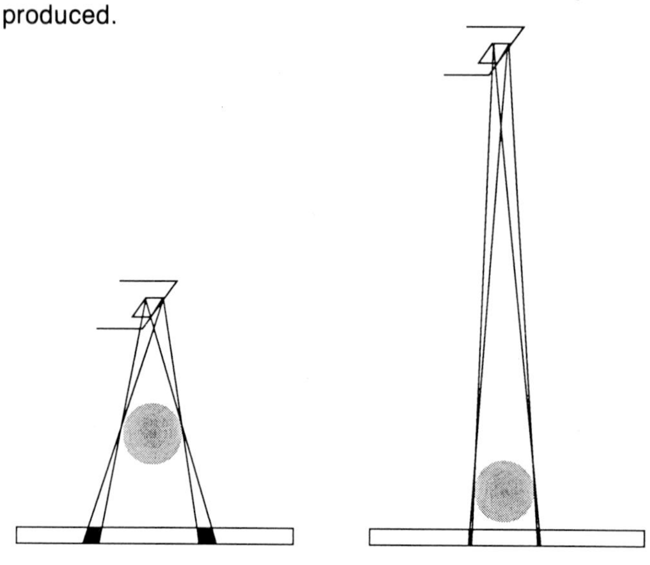

Figure 7-5. Object-to-film distance changes as they affect image sharpness. Shaded area represents edges of the anatomical structure, ideally as small as possible. Rectangular area in the tube represents the actual focal spot from which x-rays are produced.

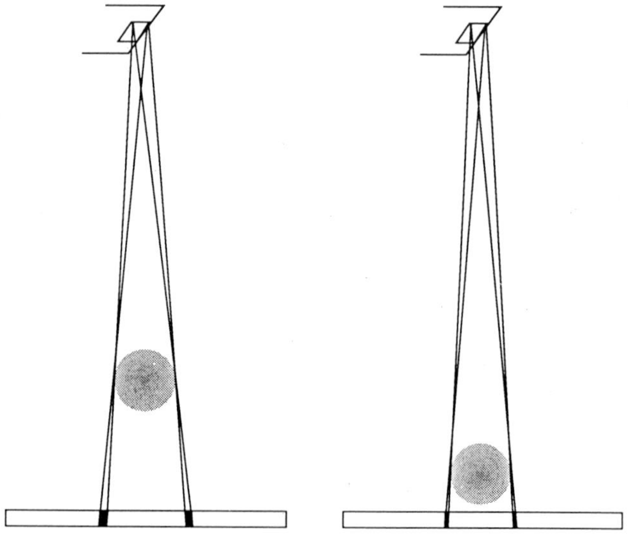

›ssesses more image sharpness than the other. Which side has ›ore detail than the other? Answer: The side closest to the film.

›igure 7-6. Radiographic effect of object-to-film distance. › PA Hand Radiographs.)
›e most noticeable effect is the size magnification caused by an increase › OFD. Note also the increase in geometric unsharpness in the radiograph ›th the longer OFD.

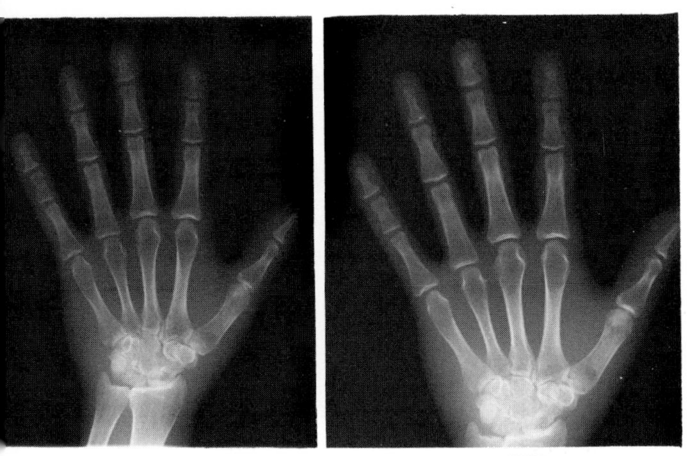

›ze and Shape Distortion

Distance affects image size, and in fact, is the critical factor › determining how much size distortion will be present on the ›diograph. As a general rule, magnification is undesirable and ›ould be minimized. Toward this end, FFD should be as long › is practical. Target-to-film distance has an inverse relation›ip to size distortion. A longer FFD results in less size distor›on; a shorter FFD results in more size distortion (Figure 7-7). ›FD should be as short as possible. OFD has a direct relationship to size distortion. A longer object-to-film distance produces more size distortion; a shorter OFD produces less size distortion (Figure 7-8).

›igure 7-7. Target-to-film distance changes as they affect ›agnification. Shaded area represents image size.

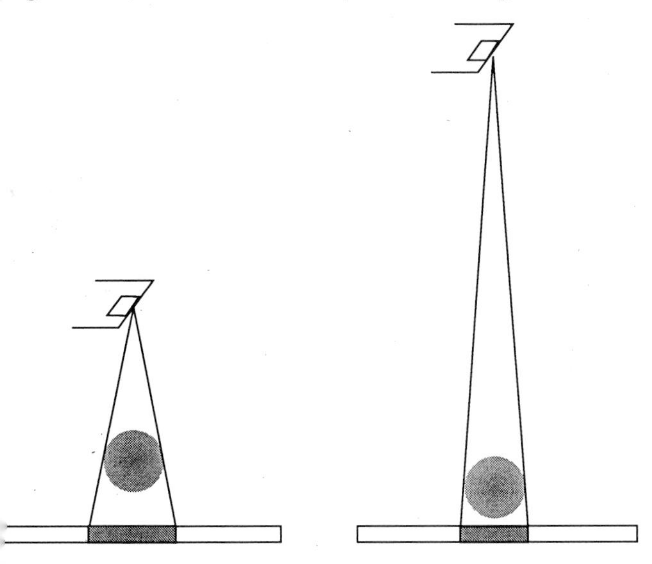

Figure 7-8. Object-to-film distance changes as they affect magnification (or size distortion). Shaded area represents image size.

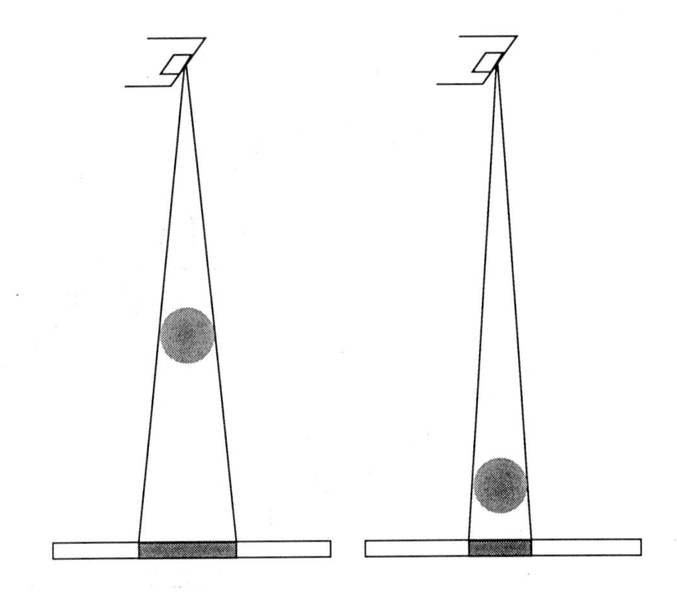

SUMMARY

An understanding of focal film distance and object-to-film distance are both important because of the related roles that each has on several aspects of radiographic quality. Focal film distance is the distance from the tungsten target of the x-ray tube to the film. Object-to-film distance is the distance from the body part to the film. FFD affects radiographic density, recorded detail, and size distortion. FFD has an inverse relationship on radiographic density, a direct relationship to recorded detail, and an inverse relationship to size distortion. Since detail is desirable and size distortion is not desirable, a longer FFD is preferable in those situations where a choice is available. To regulate density, the mAs-Distance formula must be used.

Object-to-film distance affects radiographic density, radiographic contrast, recorded detail, and size distortion. Because of its numerous effects, unintentional introduction of OFD could have serious consequences. OFD has an inverse relationship to radiographic density, a direct relationship to radiographic contrast, an inverse relationship to recorded detail, and a direct relationship to size distortion.

The intentional introduction of OFD is part of the mechanics of creating an ''air-gap'' technique, which is used to reduce the amount of scatter radiation reaching the x-ray film without resorting to a grid. At the same time the OFD is increased, FFD is increased (to offset detail and magnification problems), and the mAs is recalculated to restore the original film density.

The radiographic effects of distance are summarized in Table 7-1. Only through proper management of FFD and OFD can quality radiographic images be obtained.

TABLE 7-1. **Summary of the Effects of Distances on Radiographic Quality**

	FFD	OFD
IMAGE VISIBILITY		
Density	–	–
Contrast	0	+
IMAGE FORM		
Recorded Detail	+	–
Size Distortion	–	+
Shape Distortion	0	0

+ = Direct Relationship to Image Characteristic
- = Inverse Relationship to Image Characteristic
0 = No Effect on Image Characteristic

STUDY QUESTIONS

1. Define the terms focal film distance and object-to-film distance.

2. For a given technique, x-ray intensity is measured at 200 milliRoentgens (mR) at a distance of 36″. If no other adjustments are made to machine factors, what will be the intensity at 72″?

3. If 35 mAs is used to produce a radiograph at a 40″ FFD, what mAs will be necessary to produce the equivalent film at 50″?

4. If focal film distance is increased, what, if anything will happen to radiographic density?

5. If object-to-film distance is decreased, what, if anything will happen to radiographic density?

6. If focal film distance is increased, what, if anything will happen to radiographic contrast?

7. If focal film distance is decreased, what, if anything will happen to recorded detail?

8. If object-to-film distance is increased, what, if anything will happen to recorded detail?

9. If focal film distance is decreased, what, if anything will happen to size distortion; to shape distortion?

10. If object-to-film distance is increased, what, if anything will happen to size distortion; to shape distortion?

11. Given choices of a 10″, 40″, or 50″ FFD, identify the one that would produce a radiograph with: a) the greatest density; b) the least density; c) the most recorded detail d) the least recorded detail; e) the most size distortion; f) the least size distortion.

12. Given choices of a 0″, 2″, or 4″ OFD, identify the one that would produce a radiograph with: a) the greatest density; b) the least density; c) the most recorded detail; d) the least recorded detail; e) the most size distortion; f) the least size distortion.

BIBLIOGRAPHY

1. Bushong, Stewart C. *Radiologic Science for Technologists*, 2nd ed. St. Louis: C. V. Mosby Company, 1980.
2. Donohue, Daniel P. *An Analysis of Radiographic Quality*. Baltimore: University Park Press, 1980.
3. Gould, R. G., and J. Hale. Control of Scattered Radiation by Air Gap Techniques: Applications to Chest Radiography. *American Journal of Roentgenology*, 122:109, 1974.
4. Morgan, James A. *The Art and Science of Medical Radiography*, 5th ed. St. Louis: The Catholic Hospital Association, 1978.
5. Selman, Joseph. *The Fundamentals of X-Ray and Radium Physics*, 5th ed. Springfield: Charles C. Thomas, 1972.
6. *The Fundamentals of Radiography*. Rochester: Eastman Kodak Company, 1980.

Chapter 8

The Effect of Focal Spot Size on Radiographic Quality

Chapter Outline

1. Learning Objectives
2. Introduction
3. The Role of Focal Spots in Radiography
   A. X-Ray Production
   B. Focal Spot Structure
   C. Focal Spot Measuring Tools
   D. Focal Spot Control
4. The Effects of Focal Spot Size on Image Visibility
   A. Radiographic Density
   B. Radiographic Contrast
5. The Effects of Focal Spot Size on Image Formation
   A. Recorded Detail
   B. Size Distortion
   C. Shape Distortion
6. Summary
7. Study Questions
8. Bibliography

Learning Objectives

Upon completion of this chapter, the student should be able to:

1. List the names of four focal spots and identify their locations within the x-ray tube.
2. Explain the Line Focus Principle.
3. Give the range in degrees of the target angles used in rotating anode x-ray tubes, the most widely used angle for routine radiography, and the optimum angle for angiography.
4. Describe the relationship between anode angle and x-ray intensity.
5. Describe the "heel effect" and explain its significances to radiography.
6. List the cross-sectional range in millimeters of the optical focal spot sizes used in medical radiography.
7. Describe the effect of focal spot size on radiographic density, contrast, recorded detail, and distortion.

Introduction

The focal spot of the x-ray tube is an important radiographic tool for the radiographer in the control of recorded definition of image details. As a result, focal spot sizes require routine inspection to determine if focal spot measurements are remaining within the specifications of the x-ray tube manufacturers. Radiographers that understand the importance of selecting the correct focal spot size of dual focused x-ray tubes based on exposure requirements have the capacity for creating radiographic images with minimal geometric unsharpness.

The Role of Focal Spots in Radiography

X-Ray Production

When cathode electrons impinge upon the focal spot of the beveled anode surface, x-rays are produced. With adequate x-ray tube shielding only those x-rays that are emitted from the focal spot are allowed to emerge from the tube housing port and are used for the radiographic process. Target angles for rotating anode tubes range from about 7 to 20 degrees for radiographic work at 100 cm FFD using 35 cm × 43 cm x-ray film. The most widely used target angle is 12 degrees. For angiographic work with 35 cm × 35 cm film at 100 cm FFD, the optimum angle is 10 degrees which increases loading 20 percent with a corresponding decrease in exposure time (Seeram: 1985).

At decreased exposure times, movement of contrast media through the vascular system of a patient can be more instantaneously stopped. At reduced (steeper) anode angles, the emerging x-ray beam becomes more non-uniform since there is additional anode target material from which the emerging x-ray beam must pass. A non-uniformity of x-ray intensity exists within the diverging x-ray beam, regardless of the anode angle. This condition is known as the "heel" effect; the area of greatest x-ray intensity emerging from the tube port is under the cathode portion of the x-ray tube (Figure 8-1). Radiographers are encouraged to place ascending thicker anatomical parts in this area in order to take advantage of this intensity difference that increases with steeper anode angles.

Focal Spot Structure

The target area of the anode upon which the cathode electrons

Figure 8-1. Location of focal spot and heel effect during x-ray production. (Note that the greatest level of x-ray intensity is under the cathode side of the x-ray tube.)

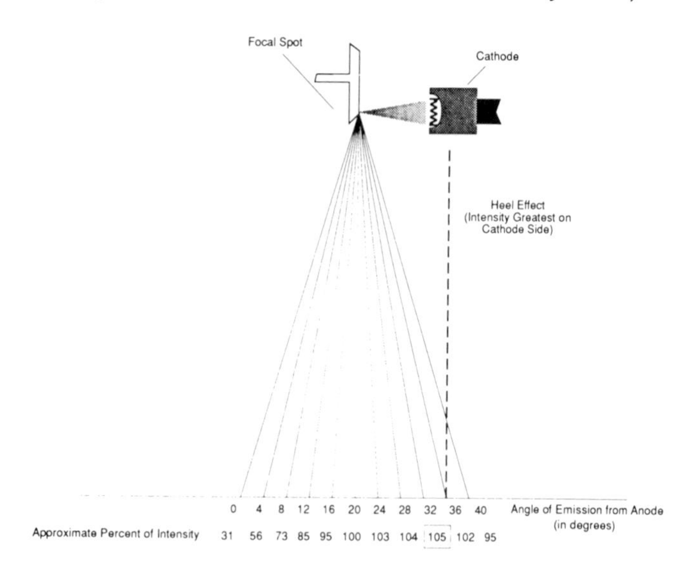

Figure 8-2. Example of Line Focus Principle. Where AB intersect anode angle defines the area of Actual or Electronic Focal Spot. This is where electron stream from cathode hits anode. Where lines CD intersect anode defines Optical or Effective Focal Spot. Steeper anode angles generate smaller Optical (Effective) Focal Spots, (EF). Likewise, smaller cathode electron streams will generate smaller Optical (effective) Focal Spots, (GH).

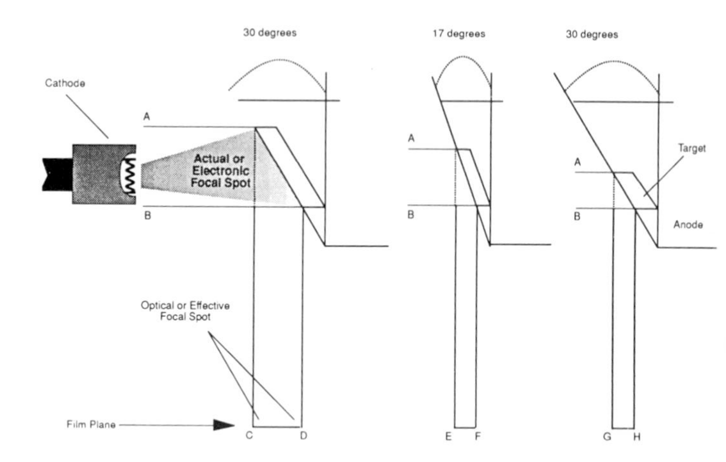

fall and from which x-rays emerge is called the focus or focal spot (Goodwin et. al.: 1970). The actual area of bombardment by the cathode electrons on the target is called the electronic focal spot (Jacobi and Paris: 1977). This can be referred to as the actual focal spot (Selman: 1976). The projected area of the electronic focal spot, which is at right angles to the electronic beam, is called the optical focal spot (Jacobi and Paris: 1977). This is also referred to as the effective focal spot (Selman: 1976). The optical focal spot is what the manufacturer cites as the focal spot size of a given x-ray tube. The optical focal spot, which is projected onto the film, is smaller than the electronic focal spot due to the angle of the anode and the actual area of electron bombardment on the anode target. This is the concept behind the Line Focus Principle as developed by Goetze (Figure 8-2). The optical focal spots of most x-ray tubes used in medical radiography have cross sections ranging from 0.3 to 2.0 millimeters.

Focal Spot Measuring Tools

Measurement of the focal spot is critical to determine whether the focal spot size is within the recommended limits for the given nominal size provided by the manufacturer. Focal spots that exceed the specifications of the tube manufacturer can cause problems with respect to the recorded definition of image details. The size of optical focal spots may be measured by three tools. The pin-hole method is an excellent way to demonstrate measurement of optical focal spot size to students. The accuracy of the pin-hole method, however, is less than that of the star-test pattern or the focal spot resolution test tool .

Focal Spot Control

The size of optical focal spot is controlled by the size of the electronic focal spot for a given anode angle. Control of the size of the electronic focal spot is accomplished by selection of the size of the cathode tungsten filament wire which is the apparatus responsible for thermionic emission of electrons. These electrons are focused onto the anode target by the molybdenum focusing cup of the cathode once the kilovoltage is applied across the x-ray tube gap. In focus biased x-ray tubes the molybdenum focusing cup is electrically negative along with the cathode filament wire. This compresses the electron beam around its circumference. Thus the electron stream bombards a smaller anode area and produces a smaller electronic focus that is more circular than square (Figure 8-3).

Two cathode filament wires, small and large, give the radiographer a choice of focal spot sizes on today's x-ray tubes (Figure 8-4). The radiographer manipulates this choice from the milliampere station settings on the x-ray control panel (Figure 8-5). Correct application of mA stations at the x-ray machine control panel and their corresponding focal spot settings may impact significantly on the geometric quality of the radiographic image.

X-ray tubes are designed such that focal spot size is directly related to the desired milliamperage setting. This is understandable since higher mA exposure settings require a significantly greater electron cloud (thermionic emission) compared to lower mA settings and higher mA settings will generate a larger area of electron bombardment on the anode target angle. On dual filament x-ray tubes, a small filament wire corresponds to lower mA settings since the amount of electrons boiled off the filament wires is lower. Likewise, a larger filament wire corresponds to higher mA settings. If these restrictions were not electronically built into the design of x-ray tubes, a high mA requirement from a small cathode filament wire would quickly result in rapid heat exhaustion or breakage under higher kV conditions. Because of this, smaller focal spots are engineered to be used with lower mA requirements while larger filament wires are used with higher mA requirements.

The Effects of Focal Spot Size on Image Visibility

Radiographic Density

As was previously discussed, the desired mA setting will influence the corresponding filament size that may be used for a given radiographic procedure. One might think that since there is a relationship between the mA setting and the size of filament wire used there is a relationship between the effect of focal spot size on radiation production and radiographic density. This is a misconception since even at low mA settings and corresponding small focal spot sizes a great amount of radiation may be produced even when long exposure times are used. Conversely, with high mA settings very low radiation intensities may be reached when exposure time settings are short. Therefore, the focal spot size does not affect the radiation output from x-ray tubes and the subsequent radiographic density.

Radiographic Contrast

The size of the focal spot does not affect the penetration power of radiation generated from x-ray tubes and, therefore, does not affect radiographic contrast. This is primarily the function of kilovoltage. However, because of a given target angle, there will always exist an intensity variation associated with the heel effect. Only when the size of focal spot is decreased by a decrease in target angle will the radiation intensity associated with the heel effect increase. However, radiographers have no control over the target angles of the x-ray tubes they operate. On the other hand, radiographers who place ascending thicker parts under the cathode end of the x-ray tube can take advantage of the increased radiation intensity which will provide greater radiographic density to thicker image details. This will assist in generating a longer scale of contrast. However, radiographers have no control over contrast based on the given focal spot size(s) of an x-ray tube, other than taking advantage of the heel effect which is dependent upon anode target angle and not focal spot size.

The Effects of Focal Spot Size on Image Formation

Recorded Detail

Since x-rays are generated from a "point source" on the electronic focal spot, the smaller that point, the smaller the penumbral effect at the point of image formation. In other words, a smaller filament wire selection on the cathode will result in a smaller electronic and optical focal spot. This will cause a smaller penumbral effect since the x-ray photons are in a more vertical relationship to the anatomy being radiographed. Consequently, there is better recorded detail from sharper structural lines (Figure 8-6).

The size of the focal spot may change insignificantly as a result of changes in mA and kV for a given cathode filament wire and a condition known as blooming occurs in which higher mA settings generate larger focal spots compared to lower mA settings. However, this has implications mainly for focal spot measuring instead of effects on the quality of the recorded definition of image details.

Figure 8-3. Function of the cathode molybdenum focusing cup. Notice that (A) without focusing cup a larger focal spot is generated compared to (B) smaller focal spot generated with focusing cup that compresses electron stream.

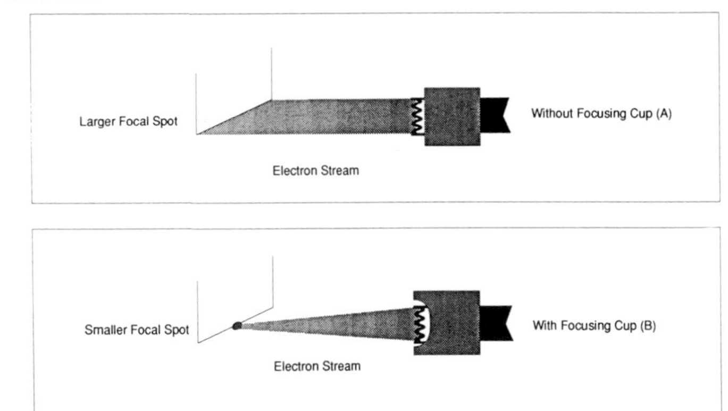

Figure 8-4. End on view of large and small cathode tungsten filament wires that are responsible for electronic focal spot.

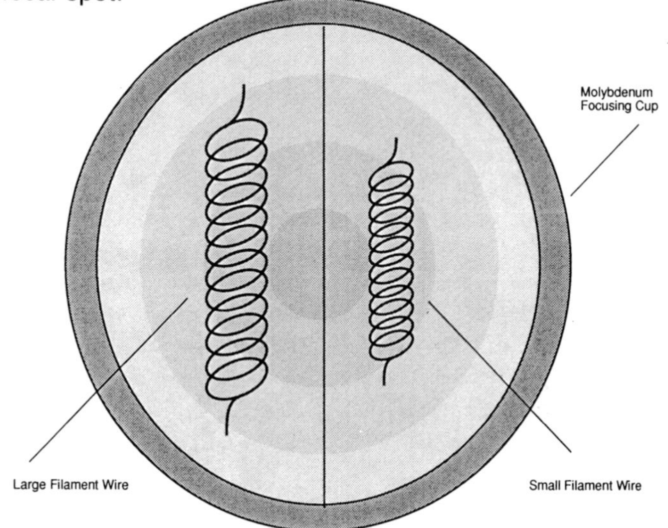

Figure 8-5. Close-up view of x-ray control panel "milliamperage, (mA), stations indicating large and small focal spot consideration.

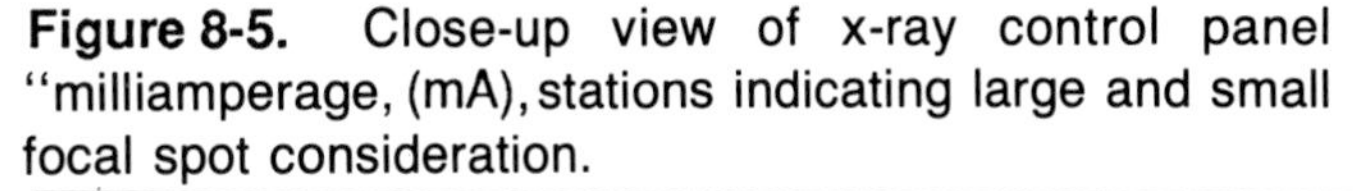

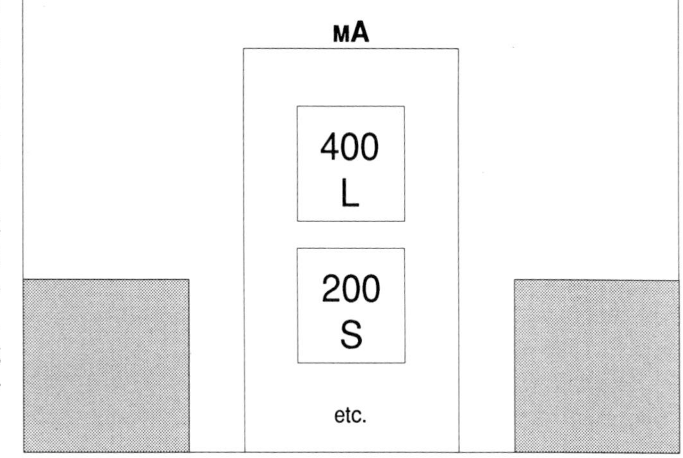

Figure 8-6. Example of the influence focal spot size has on geometric unsharpness (penumbra) of image details. For a given object film distance, larger focal spots (example A) will generate greater geometric unsharpness compared to smaller focal spot sizes (example B).

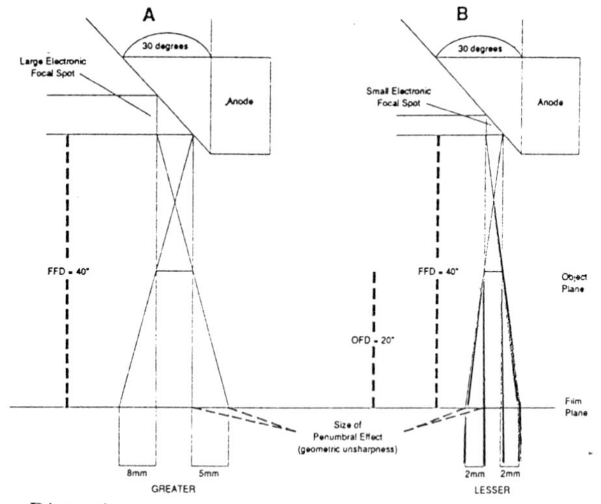

Size Distortion

Magnification radiography, also known as macroradiography, is a special radiographic technique employing the use of special "fractional" focal spot settings, for example, 0.3 mm, so that the penumbra effect is optimally minimized. Consequently, a high-quality image may be successfully magnified by increasing the object film distance. However, the variable responsible for the size distortion is distance and not focal spot size. Fractional focal spot sizes in macroradiography limit the geometric unsharpness (penumbral effect) that would otherwise result from such an increased object film distance. Changing from a small to a large focal spot setting will influence the degree of geometric unsharpness by limiting the penumbral effect, but it will not impact on the size distortion of the image details.

Shape Distortion

Obviously, if the focal spot size will not affect the size of the image details it will not distort their shape. Remember that shape distortion is caused by target, object, and film misalignment.

SUMMARY

The focal spot is that area on the anode target where cathode electrons impinge in the production of x-rays. The most widely used anode angle is 12 degrees. Radiographers are encouraged to take advantage of the anode "heel effect" by placing ascending thicker anatomical parts under the cathode side of the x-ray tube where x-rays emerging from the anode angle is most intense.

The area of electron bombardment on the anode defines the actual or electronic focal spot while the projected area at right angles to the electron beam is referred to as the effective or optical focal spot. Optical focal spot size of modern x-ray tubes range from 0.3 mm to 2.0 mm. The three tools that may be used to measure the optical focal spot size are the pinhole, star-test pattern, or focal spot resolution test tool. The size of the optical focal spot is controlled by the size of the electronic focal spot. This is controlled by filament wire selection of the cathode a the control panel.

Focal spot size has no effect on radiographic density, contras motion or screen unsharpness, or size and shape distortior However, it does impact on geometric unsharpness of imag details. Smaller focal spot sizes will generate correspondingl smaller penumbral effects which improves recorded definitio of image details. Table 8-1, provides a summary of the effect of focal spot size on radiographic quality.

TABLE 8-1. **Summary of the Effects of Focal Spot Siz on Radiographic Quality**

IMAGE VISIBILITY	
Radiographic Density	C
Radiographic Contrast	C
IMAGE FORM	
Recorded Detail	–
Size Distortion	0
Shape Distortion	0

\+ = Direct Relationship to Image Characteristic
– = Inverse Relationship to Image Characteristic
0 = No Effect on Image Characteristic

STUDY QUESTIONS

1. Give two acceptable names of the area of bombardment o chathode electrons on the target anode.
2. What is another name for electronic focal spot?
3. List three methods for measuring focal spot size of x-ray tubes.
4. Steeper anode angles result in the production of a more uniform x-ray beam. True or False?
5. What is the range of anode angulation from which most medical radiography x-ray tubes are constructed?
6. What controls the size of the optical focal spot?
7. What controls the size of the electronic focal spot?
8. What is meant by the term "focus biased x-ray tubes"?
9. Why are large and small focal spot sizes engineered to correspond to mA exposure requirements?
10. What is the effect of focal spot size on density, contrast, unsharpness, magnification, and distortion?

BIBLIOGRAPHY

1. Focal Spot Size, *Quality Assurance Procedures Manual for Diagnostic Radiology, Nuclear Medicine, Diagnostic Ultrasound*, Nuclear Associates, (Division of Victoreen, Inc.) Carle Place, New York, 1979.
2. Jacobi, Charles A., and Don Q. Paris; *Textbook of Radiologic Technology*, 6th ed.: C. V. Mosby Company, Saint Louis, Missouri, 1977.
3. Goodwin, Paul N., Edith H. Quimby, and Russell H. Morgan; *Physical Foundations of Radiology*, 4th ed.: Harper & Row, New York, 1970.
4. Selman, Joseph; *The Fundamentals of X-Ray and Radium Physics*, 5th ed: Charles C. Thomas, Springfield, Illinois, 1976.
5. Seeram, Euclid, *X-Ray Imaging Equipment*, Charles C. Thomas, Springfield, Illinois, 1985.

Chapter 9

The Effects of Filtration on Radiographic Quality

Chapter Outline

1. Learning Objectives
2. Introduction
3. The Role of Filtration in Radiography
   A. Collimator Filtration
   B. Wedge and Compensating Filters
4. The Effect of Filtration on Image Visibility
   A. Radiographic Density
   B. Radiographic Contrast
5. The Effects of Filtration on Image Formation
6. Summary
7. Study Questions

Learning Objectives

Upon completion of this chapter, the student should be able to:

1. Describe the role of filters in radiography.
2. Describe inherent versus added filtration.
3. Identify those components of a collimator that filter x-radiation.
4. Describe the use and application of wedge and compensating filters.
5. Identify the effect, if any, of a change in filtration on the following:
   a. radiographic density
   b. radiographic contrast
   c. recorded detail
   d. size distortion
   e. shape distortion
6. Given different filtration amounts, identify the one that would produce a radiograph with more (or less) density, higher (or lower) contrast, longer (or shorter) scale of contrast, as specified.

Introduction

As almost every automobile owner knows, the filters on a car play a vital operating role. While the use of filtration is not going to make the x-ray unit last any longer, it is most assuredly important for several reasons and radiographers must understand its effects.

Consider the effects of filtration changes on the radiograph. The radiographer will not be frequently changing filtration quantities; however, understanding the effects that these changes would have is still important, especially when the radiation health physicist has determined that filtration quantities are to be changed for some or all of the x-ray units. There are required minimums, but, that is not to say that an x-ray unit cannot have more, and indeed, a few do.

The Role of Filtration in Radiography

The x-ray beam is heterogeneous, meaning that the x-rays produced from the tube at any given instant are not the same wave length. Sometimes the term ''polychromatic'' is used, which, in relationship to invisible light, is a function of the wave length. Both terms then would seem to be legitimate substitutes for each other. The term heterogeneous will be used in this text because the ''color'' of invisible x-rays is difficult to conceptualize. Some x-rays have a very short wave length with greater penetrating power and some have a long wave length with less penetrating power. The average penetrating power is controlled by kVp. At a higher kVp, the average penetrating power is greater. This does not alter the fact that there are wave lengths in every x-ray beam that do not possess sufficient energy to penetrate the body part that is being radiographed, but these long wave length photons contribute to patient exposure and the production of scatter photons within the patient. Both are undesirable consequences.

The obvious solution then is to eliminate those photons from the beam that will not exit through the body part. This will substantially reduce the amount of radiation the patient actually receives and improve image quality. The substance used for this purpose in diagnostic x-ray units for many years has been aluminum, partly because of its availability, ease of shaping, and low cost. All solid substances, though, to one degree or another, absorb (or attenuate) x-rays. However, aluminum is unique in that it absorbs the low-energy (long wave length) x-rays while permitting most of the high-energy (short wave length) x-rays to pass through. The net result is that with filtration, the average penetrating power of the beam is increased (the beam is hardened), while patient exposure is decreased. Students must *not* confuse this hardening effect of filtration with greater penetration of the x-ray beam through the anatomical part. Filtration does not add energy or shorten wave length.

Filtration of modern x-ray units encompasses two components, inherent and added. As the name implies, inherent filtration is that filtration which the components of the x-ray tube naturally possess. One will not see actual aluminum in the x-ray tube as the tube is removed from the tube housing. Yet, the window of the thin glass envelope accounts for some aluminum equivalent filtration. Once in the tube housing and combined with the insulating oil and plastic housing portal, these substances have the same filtration effect as approximately 0.5 millimeters (mm) of aluminum. Because it is not actual alumninum, this amount is expressed as aluminum equivalent, meaning equal to that amount of aluminum filtration. Through long-term tube use, the inherent filtration tends to increase slightly as the tungsten filament and target are vaporized and deposited on the inner surface of the glass envelope. As Table 9-1 illustrates, the majority of the filtration effect is due to the glass, with only negligible contributions by the oil and plastic portal.

For the majority of diagnostic procedures, 0.5 mm is not an adequate amount of filtration. NCRP Report Number 33 states that for diagnostic procedures using less than 50 kVp, 0.5 mm aluminum *is* adequate. For kVp settings between 50 and 70, a total of not less than 1.5 mm aluminum *shall* be used. For kVps above 70, a total of not less than 2.5 mm Al *shall* be used (Table 9-2). The wording of the regulation leaves no room for doubt. These amounts are required minimums.

Since the inherent filtration of most x-ray machine tube arrangements is not adjustable, the amount in the beam must be the amount appropriate for the highest usable kVp, that is not less than 2.5 mm Al. Total filtration would be the sum of all filtration in the beam. If the tube has an inherent filtration of 0.5 mm and the total required amount is 2.5, it is evident that 2 extra millimeters of filtration is necessary. That is exactly the purpose of added filtration. This is added to what was present by virtue of the tube and its housing. Added filtration is either in the form of actual strips of aluminum, or in some cases, is accomplished in part by the collimator (beam-restricting device). See Figure 9-1.

We have described the filter's role as one of eliminating low-energy photons. Elimination of all low-energy photons (and none of the high-energy photons) would be best; however, a few high-energy photons will be attenuated, so there are tradeoffs to consider when using more than the minimum required amount of filtration. The amount of filtration used does have an effect on radiographic quality. The effects of extra filtration on absorption as studied by Thomas Curry and his associates is presented in Table 9-2.

TABLE 9-1. **Inherent Filtration for Typical X-ray Tube.**

Absorber	Thickness (mm)	Aluminum equivalent (mm)
glass envelope	1.4	0.78
insulating oil	2.36	0.07
Bakelite window	1.02	0.05

Figure 9-1. Collimator photograph (side view).

Note that on the side of this collimator is a lever that removes or re-inserts added filtration. In this and many other collimators, inadvertently flipping this lever will prevent exposure from being taken.

Notice that 2 mm or 3 mm of aluminum absorbs the soft rays that need to be absorbed (20 kV and below). A significant amount of additional filtration would be needed to absorb the high-energy photons, but would also result in significantly decreasing the quantity of photons left to expose the film. Since affecting film density would be undesirable, most authorities conclude that there is little advantage to be gained by using a total filtration of more than 3 millimeters.

As was previously mentioned, the use of filtration is a factor

TABLE 9-2. **Percent Attenuation of Monochromatic Radiation by Various Thicknesses of Aluminum Filtration.**

Photon Energy (keV)	Photons Attenuated (%)			
	1 mm	2 mm	3 mm	10 mm
10	100	100	100	100
20	58	82	92	100
30	24	42	56	93
40	12	23	32	73
50	8	16	22	57
60	6	12	18	48
80	5	10	14	39
100	4	8	12	35

in reducing exposure to patients. Because the x-rays that are eliminated by filtration would have been absorbed in the body part, the reason for elimination is obvious. Absorbed dose results in biological alterations and potential for damage. Table 9-3 reflects that the percentage of low-energy rays eliminated dramatically increases with more filtration. Exposure times were used to compensate for the density differences that would have otherwise resulted. Skin dose was measured over the pelvis for each film. As the numbers indicate, patient skin dose was decreased 80%. The numbers reflect statistics for a specific exam and would vary slightly for other body parts with different technical factors. The end result, however, is undeniable: *the use of adequate filtration reduces patient exposure dramatically.*

Verification that sufficient filtration is present in the beam may be accomplished with the use of a Wisconsin Test Cassette testing for the half-value layer (or HVL). Since HVL may be a new term, a simple definition is in order. The half-value layer (HVL) is the thickness of a specific interposed substance that reduces the original exposure rate by half. Various quality assurance test tools in conjunction with tabular information found in NCRP Report 33 (1975) can quickly determine whether the appropriate amount of filtration is present in the beam.

There is an exam that does not require the use of as much filtration as most other diagnostic studies. That exception is routine mammography. Because of the low-energy beam needed for this examination (less than 30-40 kV), total beam filtration is permitted to be as low as 0.5 mm aluminum equivalent. A special purpose x-ray tube with a beryllium window is sometimes used for this procedure. Beryllium has a lower density than glass giving the tube an inherent filtration of approximately 0.1 mm aluminum equivalent. Because of this, beryllium window x-ray tubes are not generally used for general purpose radiography units. If the radiographer had such an x-ray machine in the clinical environment, the added filtraion should be re-inserted into the beam after the performance of each set of mammography views. Also, colimator assemblies are generally removed and replaced with flare cones or extension cylinders during the performance of mammography exams.

Collimator Filtration

As indicated in Table 9-4, different amounts of filtration are required depending upon kVp levels. There are collimators that are designed with selectable filtration stations (of usually 0, 1, 2, and 3 mm Al). Even in the zero position, however, collimator structures like the collimator portal and reflecting mirror provide some filtration. The added filtration of the collimator assembly is usually equivalent to approximately 1 mm aluminum.

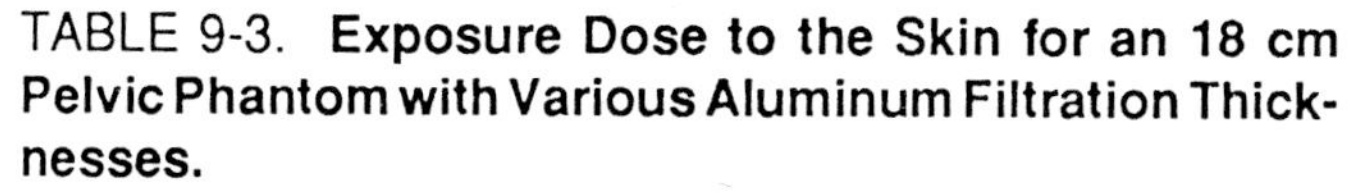

TABLE 9-3. **Exposure Dose to the Skin for an 18 cm Pelvic Phantom with Various Aluminum Filtration Thicknesses.**

	60 kVp Beam	
Aluminum Filtration (mm)	Exposure Dose To Skin (mR)	Decrease In Exposure Dose (%)
None	2380	
0.5	1850	22
1.0	1270	47
3.0	465	80

Figure 9-2. Wedge filter, used to balance densities when radiographing parts of varying thickness or tissue density.

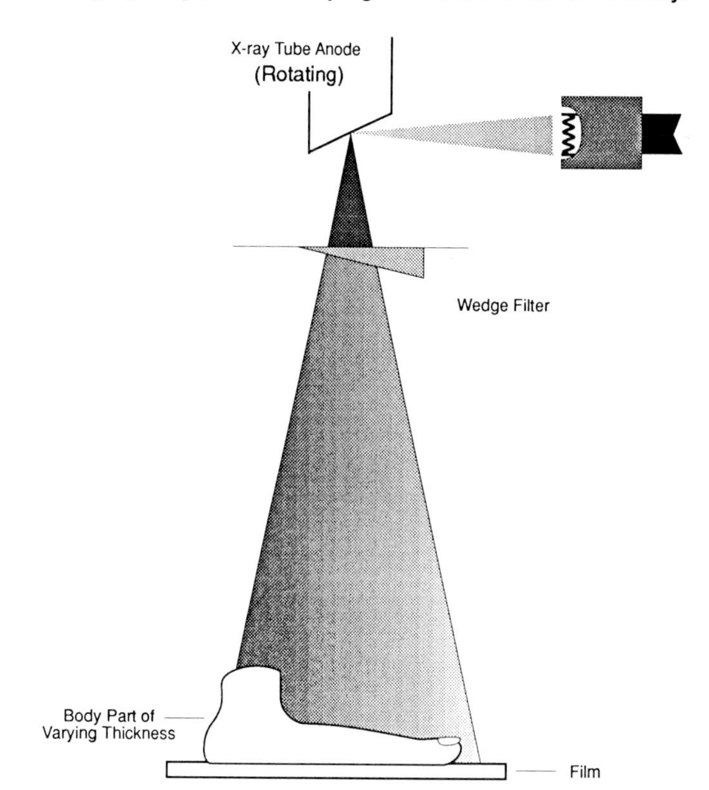

Wedge and Compensating Filters

When anatomic parts vary greatly in longitudinal thickness or tissue density, one creative technique used to obtain radiographs with uniform density is the use of wedge or compensating filters. These are sometimes called density equalization filters. As the name implies, the wedge filter is shaped like a wedge and is positioned in the beam so that the thinnest end of the wedge is over the thickest (or densest) end of the part (Figure 9-2). Less radiation is absorbed by the thin filter end, so more is available to penetrate the dense tissue. Wedge filters are extensively used in radiation therapy, but are not that often employed in diagnostic x-ray work.

TABLE 9-4. **Required Filtration.**

Operating kVp	Minimum Total Filter (Inherent plus added)
Below 50 kVp	0.5 mm aluminum
50-70 kVp	1.5 mm aluminum
Above 70 kVp	2.5 mm aluminum

Compensating filters, like wedge filters, are most commonly constructed of an acrylic type plastic, like Plexiglas, with small quantities of lead crystal (Figure 9-3). They can also be made of aluminum, brass, copper, and molding clay with kneaded-in barium sulfate powder. They are commercially available, or can be custom-made for special applications.

Compensating filters have an indicated use in obtaining radiographs of body parts which vary greatly in thickness or tissue density. It has been noted that radiographs of the spine in scoliosis studies could be improved with a more uniform exposure. For alignment purposes, compensating filters made of clear plastic mounted to the underside of the collimator are extremely easy to use. When compensating or wedge filters are used, exposure factors must be adjusted to deliver sufficient x-ray intensity for optimum visualization in the densest area of the part under consideration.

The Effect of Filtration on Image Visibility

Radiographic Density

Obviously, filtration affects the quality of the x-ray beam. With less filtration the beam is softer and with more filtration the beam is harder. There is also an effect on the intensity, or quantity, of the beam. With less filtration, there are more rays getting to the patient and, therefore, to the film. Part of the radiation that reaches the film is fog caused by the scattered interaction of the many low-energy x-rays and atoms of the body part being radiographed. It is clear, therefore, that filtration has an inverse relationship to density: more filtration decreases density while less filtration increases density. With appropriate filtration and the resulting reduction in soft rays and the accompanying fog, not as many x-rays would seem as likely to reach the film. Figure 9-4 substantiates this. Both radiographs use the identical machine factors (60 kVp, 10 mAs, 40″ FFD, non-grid). The only difference is that in the view on the right, added filtration was removed. Because radiographic density has to be maintained, the x-ray machine factors usually will have to be increased slightly if extra filtration is added. Even with this slight increase, however, the patient is still receiving less radiation than if the low-energy x-rays had not been removed.

Figure 9-3. Compensation filter, used for same purpose as wedge filter, but manufactured in different shapes for a variety of applications to various body parts.

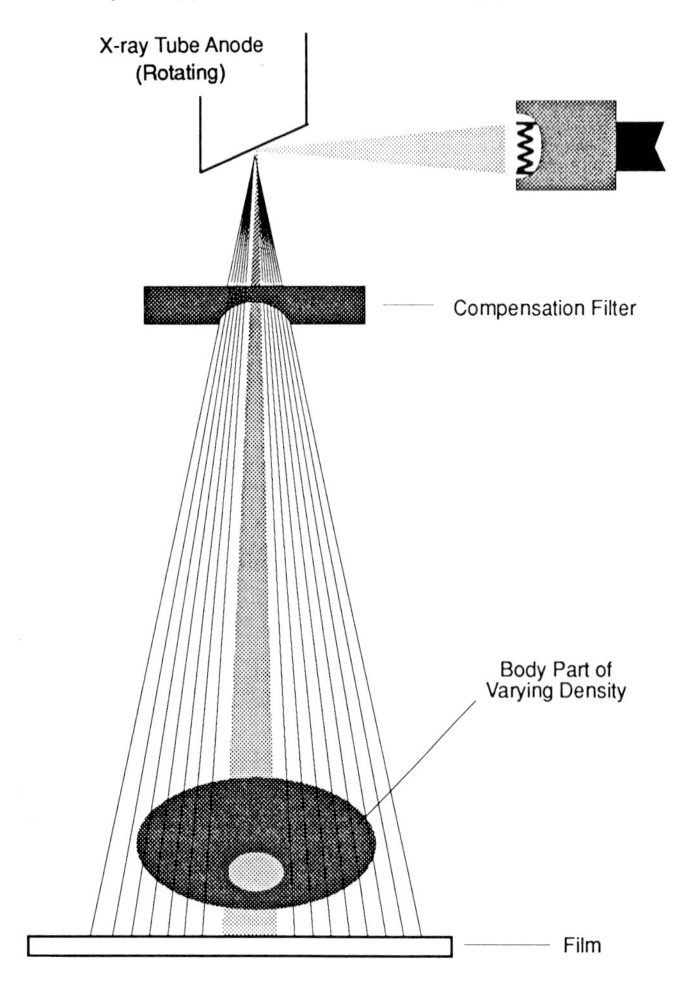

Radiographic Contrast

Since the effect of filtration is to eliminate the soft, low-energy x-rays, the x-ray beam will become more homogeneous as filtration is increased. The effect of an homogeneous beam on different tissues will be to reduce the radiographic differences between shades. Contrast decreases as filtration increases. With less filtration contrast increases. The relationship between filtration and contrast is, therefore, inverse.

In describing the scale of contrast, more filtration results in a longer scale of contrast while less filtration results in a shorter scale of contrast. Note again Figure 9-4 and compare the black versus white appearance of the skull radiograph taken without added filtration. While this film may be more pleasing to the eye, it is not necessarily the most diagnostic, and patient exposure was considerably higher.

The Effects of Filters on Image Formation

Filtration does not have an effect on either the geometric properties of recorded detail or size and shape distortion of image details. Changing the filtration may make a slight difference in the visibility of the image but not in its geometric properties.

SUMMARY

This chapter has discussed the role of filtration in reduction of patient exposure. This is accomplished by absorption of the shallow penetration x-rays produced in the x-ray tube. The resulting beam is harder, that is, the average penetrating power is higher. The filtration material of choice for diagnostic x-ray is aluminum. The x-ray tube itself absorbs some x-rays. A combination of the x-ray tube glass and the oil and air within the x-ray tube housing are referred to as the inherent filtration and absorb 0.5 mm aluminum equivalent. Actual strips of aluminum are added to bring the total filtration of an x-ray unit to a minimum of not less than 2.5 mm of aluminum or alumnimum equivalent. In some units the collimator itself becomes a part of this added filtration, because of its reflecting mirror and portal. Wedge and

Figure 9-4. Radiographs Without/With Filtration.

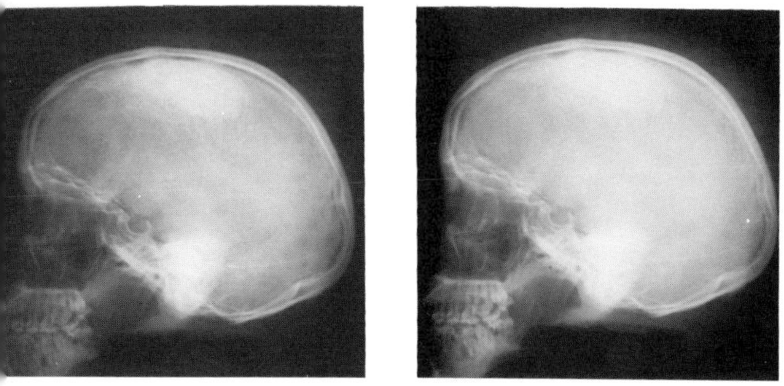

The lateral skull radiograph on the left was taken without added filtration. With identical x-ray machine factors and 4 mm aluminum added filtration, the view on the right exhibits less radiographic density and contrast.

Compensating filters have been described as methods to equalize film exposure under body parts of varying thicknesses. These are properly used by putting the thinnest end of the filter under the densest end of the part.

Radiographic density and contrast are affected by filtration. Since filtration absorbs many of the primary x-rays that would produce scatter x-radiation, the radiograph will have less density when more filtration is used. Contrast is affected similarly. Since as filtration increases, the beam becomes more homogeneous and radiographic contrast is reduced. As contrast is reduced the scale of contrast is lengthened. Filtration, therefore, affects contrast inversely and the scale of contrast directly. The radiographic effects of filtration are summarized in Table 9-5.

TABLE 9-5. **Summary of the Effects of Filtration on Radiographic Quality**

IMAGE VISIBILITY	
Density	–
Contrast	–
IMAGE FORM	
Recorded Detail	0
Size Distortion	0
Shape Distortion	0

+ = Direct Relationship
– = Inverse Relationship
0 = No Effect

STUDY QUESTIONS

1. Describe the role of filters in radiography.
2. Describe inherent versus added filtration.
3. Identify those components of a collimator that filter x-radiation.
4. Describe an application and correct use of a wedge or compensating filter.
5. If the filtration is changed from 2 to 3 millimeters, what if anything will happen to radiographic density; to contrast; to the scale of contrast?
6. If the filtration is changed from 4 to 3 millimeters, what if anything will happen to radiographic density; to recorded detail; to shape distortion; to size distortion?
7. In two identical radiographic situations different only in the amount of filtration used (one unit has 2 mm aluminum added—the other has 3 mm), which would produce a film with the most density? Which would have the highest contrast?

Chapter 10

The Effects of Beam Restrictors on Radiographic Quality

Chapter Outline

1. Learning Objectives
2. Introduction
3. The Role of Beam Restrictors in Radiography
4. The Effect of Beam Restrictors on Image Visibility
   A. Radiographic Density
   B. Radiographic Contrast
5. The Effect of Beam Restrictors on Image Formation
   A. Recorded Detail
   B. Size and Shape Distortion
6. Summary
7. Study Questions
8. Bibliography

Learning Objectives

Upon completion of this chapter, the student should be able to:

1. Describe the role of beam restrictors in radiography.
2. Describe the various devices used as beam-restricting devices.
3. Identify the effect, if any, given a change in the beam size on the following:
   a. radiographic density
   b. radiographic contrast
   c. recorded detail
   d. size distortion
   e. shape distortion
4. Given different beam sizes, identify the one that would produce a radiograph with more (or less) density, higher (or lower) contrast, longer (or shorter) scale of contrast, as specified.

Introduction

Many patients undergoing x-ray examination know that radiographers have a light source in that "little box" above them (correctly called a collimator) that helps in aligning the body part under the beam. In chest radiography, a few of those individuals have had the unfortunate experience of seeing much of the exposure room wall illuminated by that light during the preliminaries to a chest radiograph. If final adjustments were not made to the collimator to reduce the light field, a patient would be exposed to many more photons than the exam warranted. This chapter will explain the effects of beam-restricting devices on both patient exposure and radiography quality.

The Role of Beam-Restrictors in Radiography

Beam restrictors, as the name implies, limit the field of radiation. By reducing the amount of radiation reaching the patient, patient exposure is thereby reduced. There have been several different types of beam-restricting devices manufactured throughout the years. Historical references indicate that manufacturers and radiographers have not been as concerned about minimizing patient exposure in the past as they are today. In early years, radiation protective methods were rather unsophisticated or not universally practiced.

The first type of beam-restricting devices were pieces of lead with circular openings of different sizes that could be inserted into slots directly below the x-ray tube (Figure 10-1). These pieces of lead, called *lead diaphragms*, or *aperture diaphragms*, or more simply *diaphragms*, had some notable disadvantages. Lead is a soft, malleable material that wears and bends easily, making its "fit" not always perfect. Also, since the openings were round and the films were rectangular, radiation field size usually exceeded the size of the film. There were no guarantees to the patient, and little incentive for the radiographer to verify, that the diaphragm used was the one most appropriate for that particular size of film.

There are still instances today where diaphragms (of a significantly improved type) are used. Radiology departments which take large numbers of specialized views of the skull, will quite often have a skull radiography unit that makes use of special sets of diaphragms. Chest rooms sometimes use permanently attached diaphragms also. The manufacture of these diaphragms has significantly improved. They are now solidly made, result in very precise alignment, and assist in producing radiographs of optimal quality.

Next in the evolutionary progression of these devices were *flare cones* attached just beneath the x-ray tube held temporarily in

Figure 10-1. Lead diaphragm, one of the oldest types of beam-restricting devices. Other diaphragms, with different size or shape oepenings, could be used interchangably.

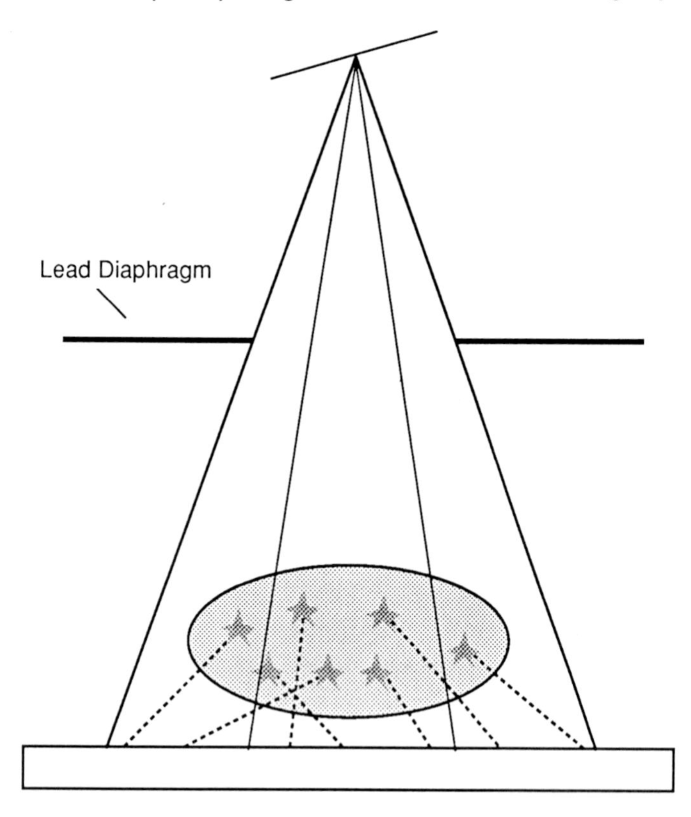

place by a "thumbscrew-type" flange (Figure 10-2). As the name implies, flare cones were megaphone-shaped devices. They came in sets with as many cones in the set as there were film sizes used in the facility. That type of attachment did give flare cones the immediate advantage of being perfectly centered under the tube. Once installed, there was still not the assurance that the tube was properly positioned over the part, that the cone was the correct one for the film being used, or even that a cone had been attached. Because cones are still commonly used in dental radiography, most people who have ever sat in the dentist's chair can remember seeing such a device protruding from the dental x-ray tube housing.

In the very same aperture of the x-ray tube housing, *extension cylinders* could also be connected. Extension cylinders, as the name implies, were "telescoping" open cannisters. Despite their physical size generally, they could produce smaller size radiation fields than even the longest, narrowest flare cone (Figure 10-3). Alignment of extension cylinders, like diaphragms and flare cones, over the patient was occasionally imprecise, primarily because there was no way to insure that all elements of the situation were in proper alignment with each other.

The more sophisticated and familiar of the beam-restricting devices is the *collimator*. This permanently mounted metal box attached to the underside of the x-ray tube housing contains at least two pairs of lead "shutters" (Figure 10-4) with a light source and relfecting mirror for indicating radiation field size and central ray alignment (Figure 10-5). The radiation field can not only match perfectly the shape and size of the film, but the internal light source also allows for a more precise alignment of the x-ray tube with the patient and film. This is done with two intersecting lines painted on the bottom of the collimator that project a "cross-hairs" pattern onto the patient, table, or film.

In August of 1974 the Department of Health and Human Resources, then HEW (which has some regulatory responsibilities over vendors of new x-ray units) began requiring new x-ray units installed after that date be equipped with "positive beam limitation." This means that the collimator will either automatically adjust to the size of the film through sensing switches and relays in the Bucky tray, or the unit will fail to take the exposure until the collimator is manually adjusted to the size of the film. Most machine manufacturers have adopted the former method. As long as the sensing devices are functional and calibration is periodically checked and corrected, these newer collimators have made the aspects of both patient positioning and radiation protection much simpler and more accurate.

There are two basic problems that occur with collimators. First, collimator light bulbs do fail, and anyone who has attempted to

Figure 10-2. Flare cone. The cone is removable and can be replaced with others in the set suitable for different size films or radiation field sizes.

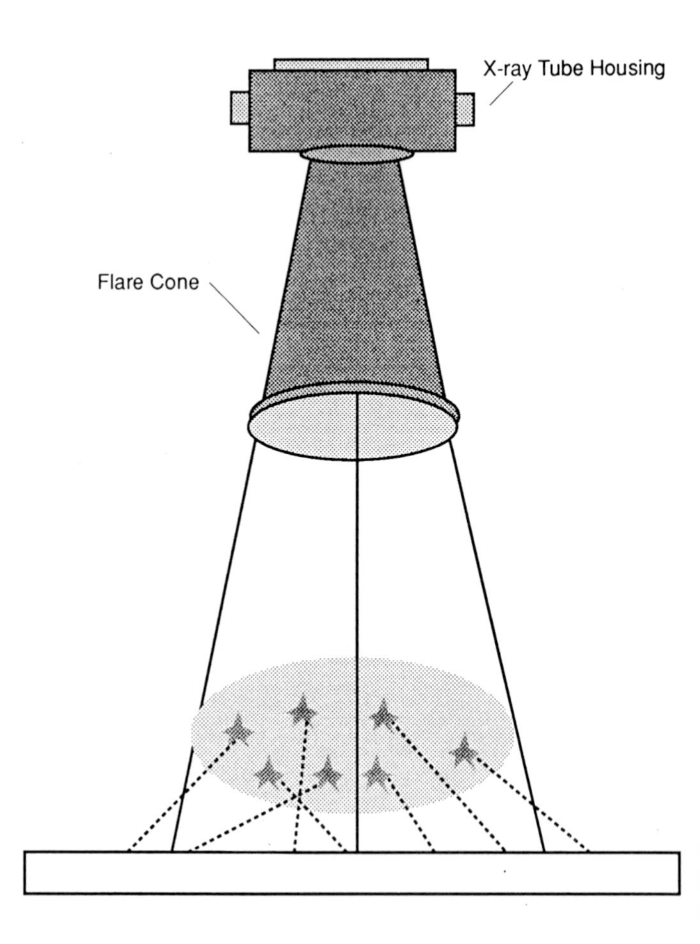

eplace one knows that installation on some collimators can be difficult. Radiographers should first acquire the necessary skills to replace a collimator light bulb before attempting replacement. Another common problem is misalignment between the light field and radiation field. There are various methods of determining congruence. Misalignment indicates that an adjustment of the light source or reflecting mirror is needed. Figure 10-6 shows a picture of a cassette that has had some paper clips bent into "L-shapes." These have been positioned into the outline of a small square or rectangle and have been taped into place with letter markers to give correct orientation of the resulting radiograph. The radiograph indicates that a minute adjustment of the reflecting mirror is necessary. Collimators are sophisticated electrical devices, and are also subject to the same kinds of malfunctions as any other piece of frequently used equipment.

The Effect of Beam Restrictors on Image Visibility

Radiographic Density

Radiographers understand that scatter radiation detracts from the quality of a radiographic image. One means of reducing the amount of scatter radiation being produced is to reduce the size of the area being irradiated. Since eliminating the production of scatter radiation will decrease the number of photons reaching the film, density decreases. Increasing the size of the radiation field by opening the collimator or using a larger cone exposes

Figure 10-3. Extension cylinder has an adjustable length. As cylinder is extended, radiation field is reduced and patient radiation exposure is decreased.

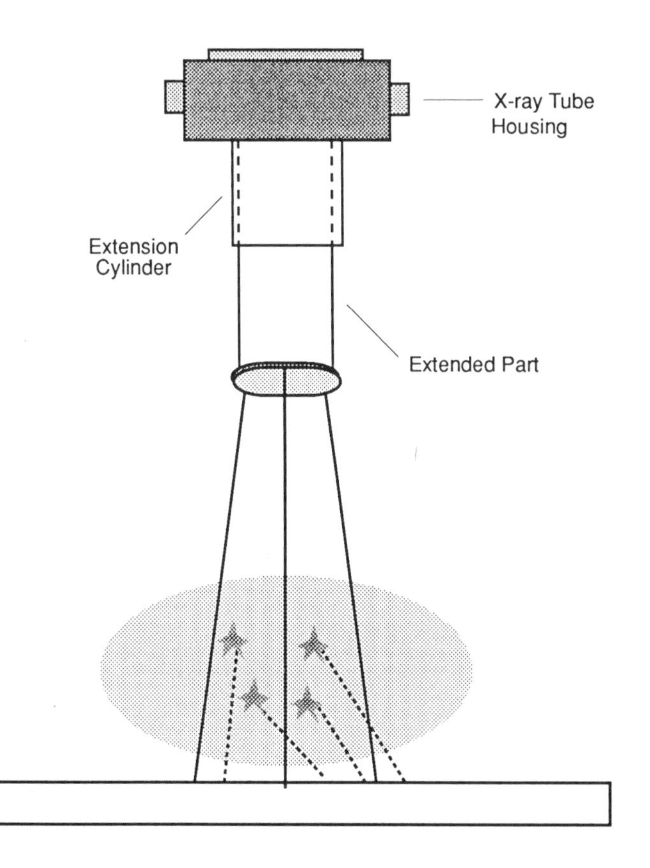

Figure 10-4. Lead Collimator Shutters (top view).

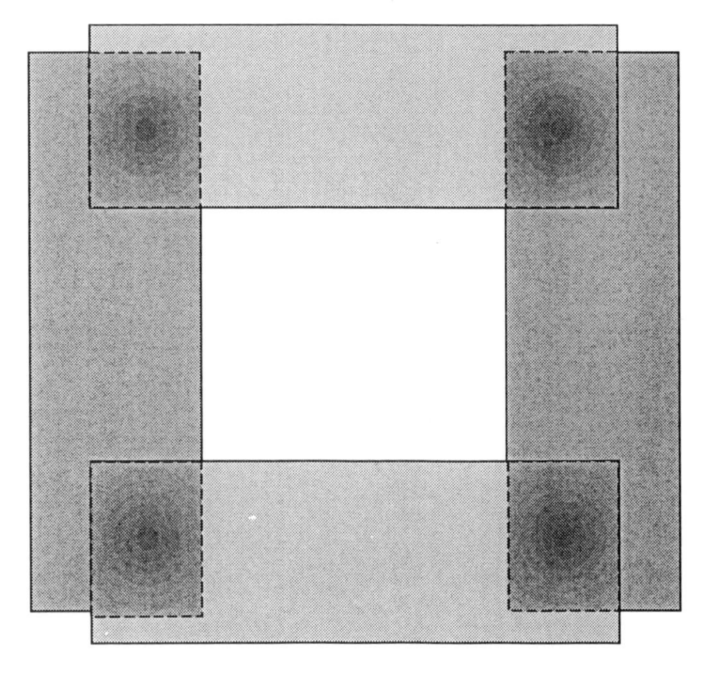

a larger area of the body part. More scatter radiation produced within the body part translates to a greater density on film. The relationship between radiation field size and density is therefore direct. A larger radiation field will produce more density even if it is slight. A smaller field will result in less density. After the radiation field reaches a size of 30 by 30 cm, the total quantity of scatter radiation (reaching the film) is near its maximum.

Beginning radiography students are sometimes surprised to learn that field size has an effect on density. An example that well illustrates this concept involves radiography of a gallbladder. The first radiograph produced in many clinical environments is usually a scout or survey film of the entire abdomen region taken on a 14 × 17-inch film. If the density is correct on this film and the gallbladder is visible, other views may be taken using smaller films and restricted fields. One may assume that because the original density is good, the same technical factors may be used on the smaller films. That radiographer is then surprised when the films are returned with less than optimal density. To offset this effect, the radiographer should increase mAs by about 60%. The exact amount may depend on patient thickness and kVp used. In Figure 10-7, notice that the second view (B), has less density than the first radiograph (A). Identical technical factors were used; hence, reducing radiation field size reduces the radiographic density. In the final film (c) the mAs was increased by 60% resulting in a density approximating the original density. An equivalent increase in kVp (about 7%) would achieve a similar radiographic density, but would produce a lower contrast, generally undesirable in gallbladder radiography.

Radiographic Contrast

Because beam size affects the quantity of radiation produced within the patient (or scatter radiation), radiographers should

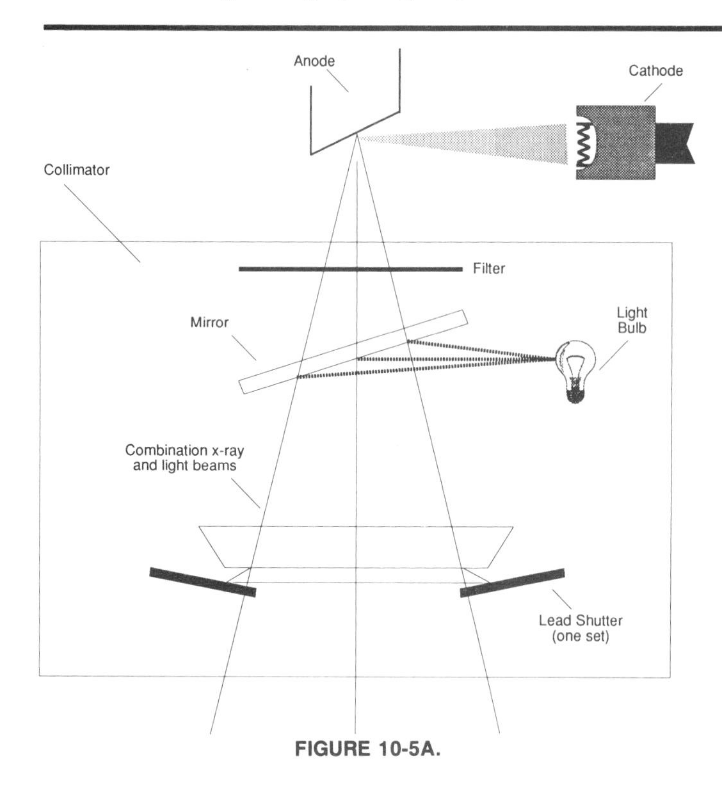

FIGURE 10-5A.

Figure 10-5. Simplified workings of colilmator are shown in 10-5A. Actual collimator photograph, front view, 10-5B shows shutter control knobs and light source switch. From the bottom of a different collimator 10-5C, note the "cross-hair" lines which will be projected to patient, table, or film for alignment purposes.

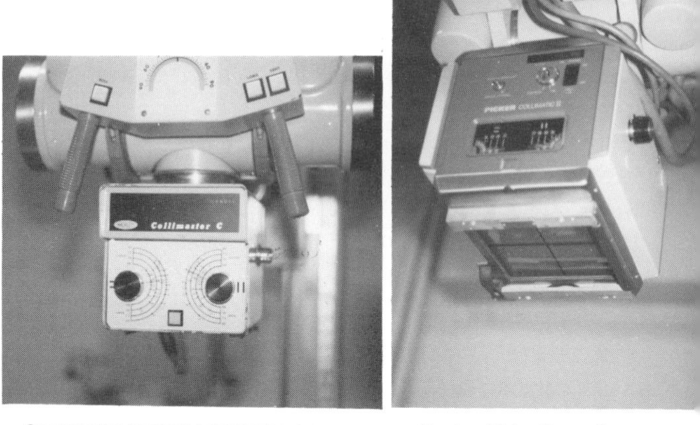

Courtesty the Machlett Laboratories, Inc.

FIGURE 10-5B.

Courtesy Picker Corporation

FIGURE 10-5C.

not be surprised that contrast is also affected. Scatter radiation produces many of the intermediate shades of gray that result in a long-scale, low-contrast radiograph. Removing the source of that scatter radiation will result in a radiograph that has a more "black-and-white" or short-scale, high-contrast appearance. The relationship between field size and subject contrast is therefore inverse. A larger field results in lower contrast; a smaller field results in higher contrast. Since cones absorb some of the radiation, small films will tend to be shorter in scale of contrast unless compensations are made in wave length by increasing kVp. In certain instances, such as gallbladder radiography, changes in conefield size are compensated for with changes in mAs, since lower contrast is generally a negative in this type of study.

Figure 10-6. Photograph 10-6A shows a cassette with paper clips and letter markers. Filmholder is positioned in light field with collimator settings adjusted to the paper clips. Radiograph 10-6B indicates that light field and radiation field are not congruent. Light field mirror needs small adjustment.

FIGURE 10-6A.

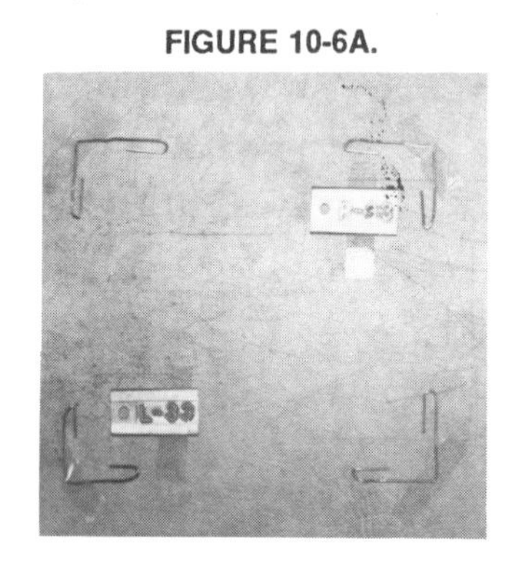

FIGURE 10-6B.

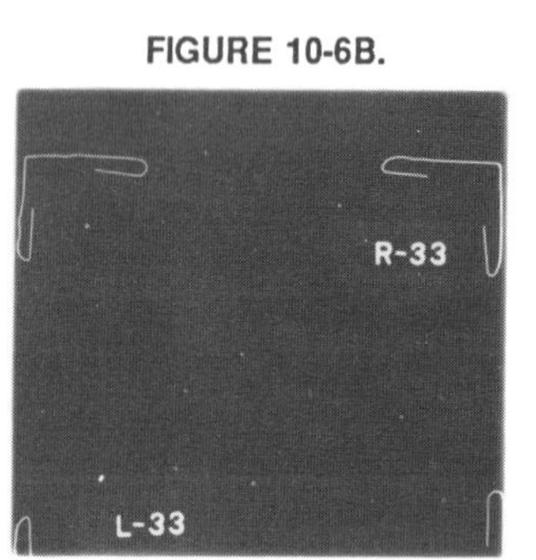

Figure 10-7. Radiograph 10-7A is a gallbladder scout film on a 14×17 filmholder with the beam size adjusted to full film coverage. Two additional views are taken on 8×10 filmholders with 6×6-inch radiation fields. Radiograph 10-7B uses original x-ray machine factors. Note the decreased density. 10-7C compensates for this density loss with a 60% increase in mAs.

FIGURE 10-7A.

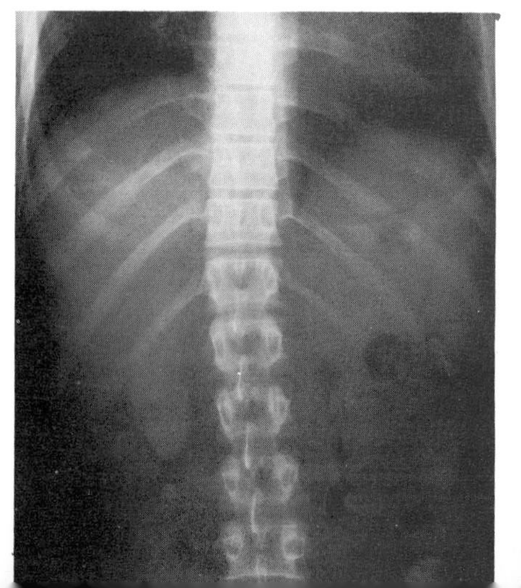

FIGURE 10-7B.

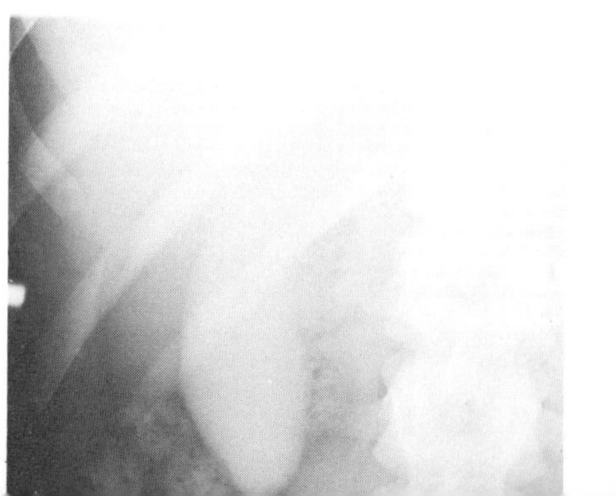

FIGURE 10-7C.

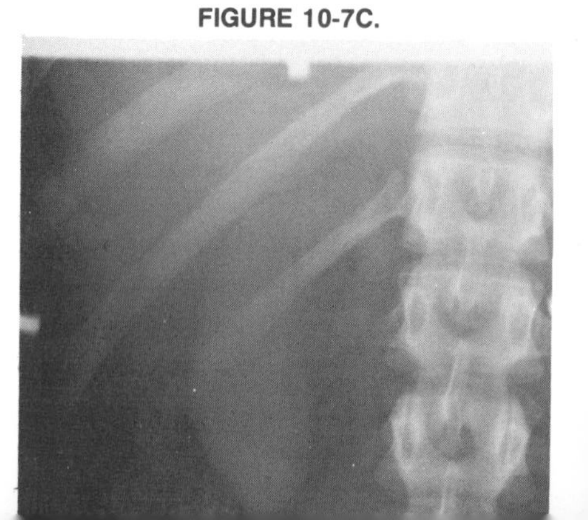

The Effect of Beam Restrictions on Image Formation

Recorded Detail

Scatter radiation and anything that allows it to be produced, or permits it to reach the radiograph, is going to have an effect on the "hiding" kind of fogging density, and some anatomical structures might be overlooked. The presence of such anatomical structures, however, is not what constitutes image sharpness. As Jacobi (1960) concluded, "radiographic sharpness (detail) may be present but obscured." Beam-restricting devices do not, therefore, have an effect on recorded detail.

Size and Shape Distortion

The use of a beam-restricting device and limited radiation field sizes does not affect the distance between the tube and film or the patient and film, nor does it affect the angulation of tube, part, or film. Therefore, *no* relationship between field size and image size or shape exists.

SUMMARY

This chapter has examined the role of beam-restricting devices in radiography. As radiation-protection devices, they play a critical role in minimizing patient exposure. Various kinds of beam-restricting devices have been used through the years including lead diaphragms, pieces of lead with usually circular openings placed directly below the x-ray tube. Lead diaphragms replaced flare cones and extension cylinders. These devices fit into a collar and were secured by thumbscrews. Precise alignment between the x-ray tube, restricting device, patient, and film, was difficult, occasionally resulting in misaligned radiographs.

The successor to these devices is the collimator, a box attached directly beneath the x-ray tube, containing at least two pairs of lead shutters with a light source and reflecting mirror to indicate radiation field size and central ray alignment. Because beam-restricting devices limit the area of exposure to the patient, there is a corresponding reduction in the quantity of radiation produced within the patient. It is this reduction in the quantity of scatter radiation which results in density and contrast changes with large adjustments in field size. As field size increases, the amount of fog reaching the film increases. This causes density to increase slightly and at the same time causes contrast to decrease slightly. Smaller field sizes decrease density and increase contrast. Field size does not affect recorded detail, size distortion, or shape distortion. The radiographic effects of beam restrictors on radiographic quality are summarized in Table 10-1.

TABLE 10-1. Summary of the Effects of Beam Size on Radiographic Quality.

IMAGE VISIBILITY	
Radiographic Density	+
Radiographic Contrast	−
IMAGE FORM	
Recorded Detail	0
Size Distortion	0
Shape Distortion	0

\+ = Direct Relationship to Image Characteristic
− = Inverse Relationship to Image Characteristic
0 = No Effect on Image Characteristic

STUDY QUESTIONS

1. What is the role of beam restrictors in radiography?
2. Describe the different types of beam-restricting devices.
3. If the beam size is changed from 14 × 17 inches to 8 × 10 inches, what, if anything, will happen to radiographic density?
4. If the beam size is changed from 8 × 10 inches to 14 × 17 inches, what, if anything, will happen to subject contrast?
5. If the beam size is changed from 8 × 10 inches to 14 × 17 inches, what if anything, will happen to the scale of contrast?
6. If the beam size is changed from 8 × 10 inches to 14 × 17 inches, what, if anything, will happen to the recorded detail; image size distortion; image shape distortion?
7. If two radiographic situations differ only in the size of radiation field (same patient, same anatomical area, same machine factors), in which size field, small or large, will there be more radiographic density?
8. If two radiographic situations differ only in the size of radiation field (same patient, same anatomical area, same machine factors), in which size field, small or large, will there be more subject contrast?
9. Two views involve the same body part (the abdomen, let's assume), same kVp, etc. They differ in that Exposure A uses a 14 × 17-inch beam at 80 mAs; Exposure B uses an 8 × 10-inch field with mAs. Which film (A or B) will have the most radiographic density? Why?

BIBLIOGRAPHY

1. Cahoon, John B. *Formulating X-Ray Techniques*, 8th ed. St. Louis: C. V. Mosby Company, 1980.
2. Curry, Thomas S., James E. Dowdey, and Robert C. Murry, Jr. *Christensen's Introduction to the Physics of Diagnostic Radiology*, 3rd ed. Philadelphia: Lea & Febiger, 1984.
3. Jacobi, Charles A. and Don Q. Paris. *Textbook of Radiologic Technology*, 1st ed. St. Louis: C. V. Mosby Company, 1960.
4. *How to Prepare an X-Ray Technic Chart*. Milwaukee: General Electric Medical Systems, 1965.

Chapter 11

The Effects of Tube-Object-Film Alignment on Radiographic Quality

Chapter Outline

1. Learning Objectives
2. Introduction
3. Tube-Object-Film Alignment in Radiography
   A. Geometric Components of Alignment
   B. Tabletop Alignment in Radiography
   C. Alignment Using the Bucky
   D. Alignment Using Floating Tabletops
   E. Horizontal Central Ray Alignment
   F. Alignments in Portable Radiography
   G. Distortion
4. Summary
5. Study Questions
6. Bibliography

Learning Objectives

Upon completion of this chapter, the student should be able to:

1. Describe the radiographic effect of improper alignment between tube, object, and film.
2. Describe the sequence of steps to be performed when using an angled central ray.
4. Describe the technique of "framing."
5. Describe shape distortion.
6. Explain why perpendicular central rays cannot always be used.
7. Explain foreshortening.

Introduction

Experienced radiographers understand the importance of proper alignment between tube, object, and film. This is very similar to the need of a photographer having his camera correctly aimed at his subject. The problem, however, is more complicated to the radiographer because there is a third component, an image receptor, which must also be correctly aligned with the "camera" and subject. A discussion of typical alignments between these components of radiographic imaging, along with common problems of achieving proper alignment follows.

Unlike previous chapters which dealt with the effects of a particular variable on radiographic quality, such as density, contrast, unsharpness, magnification and/or distortion, this chapter focuses primarily on tube-object-film alignment as a multi-faceted variable which affects the geometry of image formation. Since tube-object-film alignment is geometric and not photo-radiographic in nature, one can assume that tube-object-film alignment will not impact on the visibility of image details, such as density and contrast. Of primary concern is the accurate portrayal of anatomical structures as called for by the given radiographic projection being performed, as well as the overall aesthetic appearance of the x-ray product.

Tube-Object-Film Alignment in Radiography

Ideally, optimal alignment exists with a perpendicular central ray directed to the center of the film, or center of the film area used, in cases where more than one view is to be taken on the same film. Then the object/subject is positioned in such a way that the long axis or broad surface of the structure is parallel to the film (and therefore perpendicular to the central ray); and that the structure identified as the center of the image for any given projection is aligned under the CR and over the film midpoint.

Geometric Components of Alignment

All three components of radiographic alignment — tube, object, and film — must be geometrically correct to obtain optimum radiographic imaging. If any one of the three is not in alignment with the other two, the result will be an image that is not perfectly centered or standardly demonstrated at best, or an unsatisfactory radiograph at worst. Figure 11-1 portrays three radiographs showing: A. tube misalignment, B. patient misalignment, and C. film misalignment. All three errors can have a variety of causes and can occur in combination with another.

Tabletop Alignment in Radiography

Alignment is most easily verified when the film is in direct line of sight and can easily be seen immediately below the body part. The anatomical centering point for a give radiographic projection must be under the central ray of the primary beam. The

Figure 11-1. Radiographic Examples of Misalignment.

Film and patient correctly centered, x-ray tube 3″ off center.

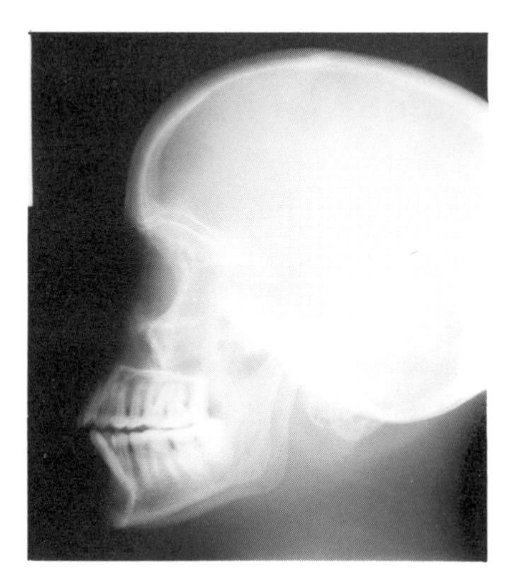

Film and x-ray tube correctly centered, patient 3″ off center.

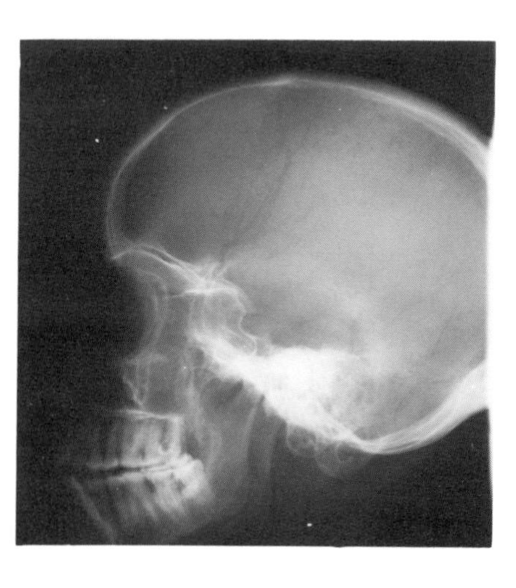

Patient and x-ray tube correctly centered, Bucky tray (with film) not completely pushed under the table.

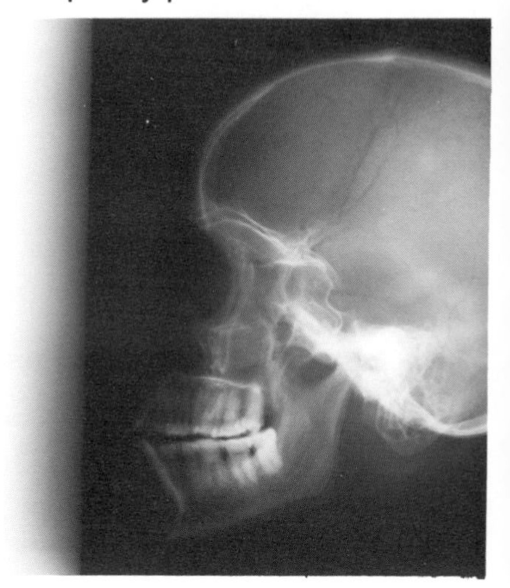

center of the part must also be directly over the center of the film holder. In the event that multiple views are to be taken on one filmholder, the part and central ray must be over the center of the film area used. The collimator's light field and "cross hairs" serve to dramatically reduce the number of errors of patient misalignment.

Alignment Using the Bucky

Precautions should also be taken to ensure that the alignment procedure is done in a logical sequence with Bucky work because direct line-of-sight is lost with the film holder. Collimator lights and light-localizing beams can be of great assistance for centering to the exact point in both the longitudinal and transverse planes of the film. Requirements of different radiographic examinations may modify the sequence of the positioning steps that a radiographer uses. (For example, using an angled central ray, centering of the ray could only be done *after* angling the tube.) If the x-ray tube is first aligned to the patient, then the film should be moved by sliding the Bucky tray to its correct position under the patient. If the tube were already correctly aligned to the patient, moving the tube again would be a mistake. Conversely, if the patient was aligned to the film, or film to the patient, then the tube becomes the final step in the alignment procedure.

Failure to completely close the film tray of the Bucky is an occasional problem. Most radiographers with experience will not make this mistake, but this possibility does exist and is usually serious enough in consequence to result in an unsatisfactory radiograph (Figure 11-1C).

Another occurrence with Bucky work is failure to secure the film properly in the tray. This can mean that the lock is not secured or that before the lock is secured, the clamps, which laterally center the film, are not pushed firmly together (Figure 11-2). Either way, if the tray is firmly closed as it should be, the film will most likely be off center to the patient and tube. Again, awareness is the key to avoiding these mistakes.

Alignment Using Floating Tabletops

One exception to the preceding statement has to do with "floating" tabletops. Locks normally holding the table in place can be released to allow the table to "float," or roll smoothly in any horizontal plane over the top of the Bucky that is underneath. This feature is very helpful with patients that are not completely ambulatory, but can present difficulties to the radiographer that does not routinely use this equipment. With "floating" tabletops there are normally "detent" locks that will

Figure 11-2. Unsecured film in Bucky. Example of one of the errors caused by carelessness. Film that is not securely locked in the Bucky tray will usually be at least off-center from the central ray.

assist the radiographer in finding the Bucky center with the x-ray tube. By depressing this "center," or sometimes called "transverse center" button and then moving the x-ray tube, the tube's movement will be halted when it is over the centerline of the Bucky. Once the tube and film are aligned to each other, the patient on this "floating" table is moved to the exact position needed under the central ray before the table is once more secured in a nonfloat position.

Horizontal Central Ray Alignment

For upright work (horizontal central rays) some radiographers find it helpful to establish a mark (if not provided by the manufacturer) on the overhead track or framework on which the x-ray tube support assembly is mounted. This mark will show the exact position of the x-ray tube stand when the tube is centered to the wall-mounted cassette holder or upright Bucky.

Alignments in Portable Radiography

With portable units, radiographers are faced with different equipment and a different set of problems. Since sliding a filmholder under a patient may obscure the film from view, the radiographer must rely on a combination of experience and knowledge to properly position the x-ray tube.

One feature that many radiographers use to assist them in positioning the tube involves the technique known as framing. *Framing* is measurement of the light field from the collimator on either side of the body part. If the light field is unequal on the two sides of the film, and the film is in the proper place, the x-ray tube needs to be moved in the direction that will make the light fields equal on both sides, or the patient needs to be moved within the light field (Figure 11-3). When using this method, it is imperative that the collimator settings be returned to their original positions for reduction of radiation exposure to the patient before the exposure is taken.

Figure 11-3. Misaligned x-ray tube (top view). Extra light field on the right indicates that the x-ray tube needs to be shifted to the left. If when centered to the patient, light borders compared to the edges of the film are still unequal, movement of the film would be indicated.

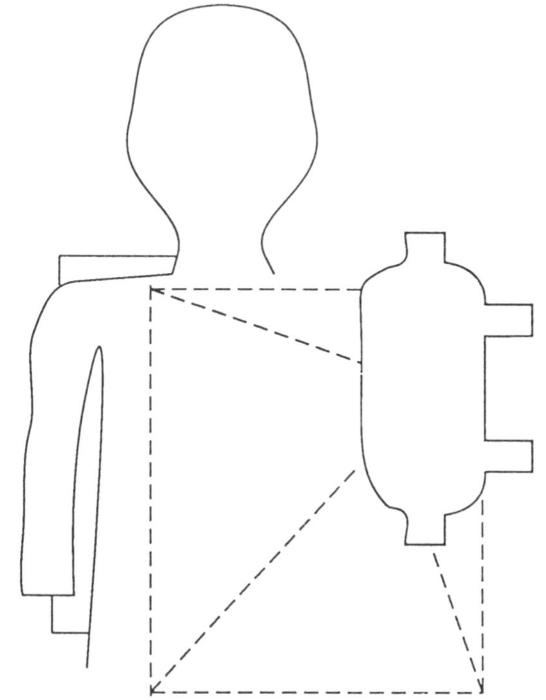

Figure 11-4. Marked filmholder borders make alignment of the film to the x-ray tube (or vice-versa) easier, especially when this film holder is partially obscured under the patient.

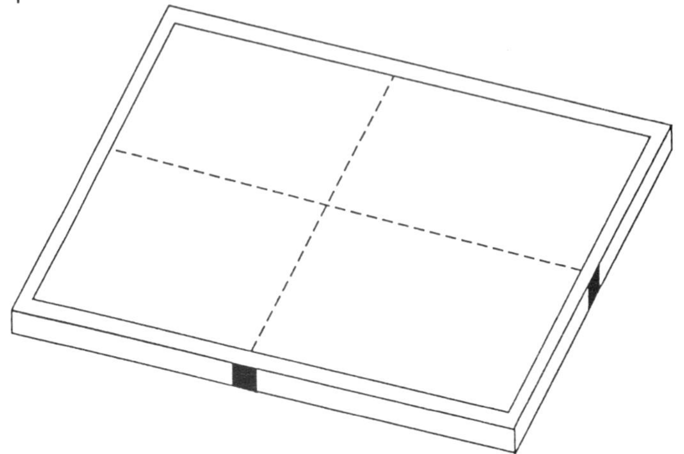

Figure 11-5. Partially obscured filmholder is a commonplace situation in portable radiography. In this instance, collimator light field with "cross-hair" markings can be used to properly align marked filmholder to the x-ray beam, or vice-versa.

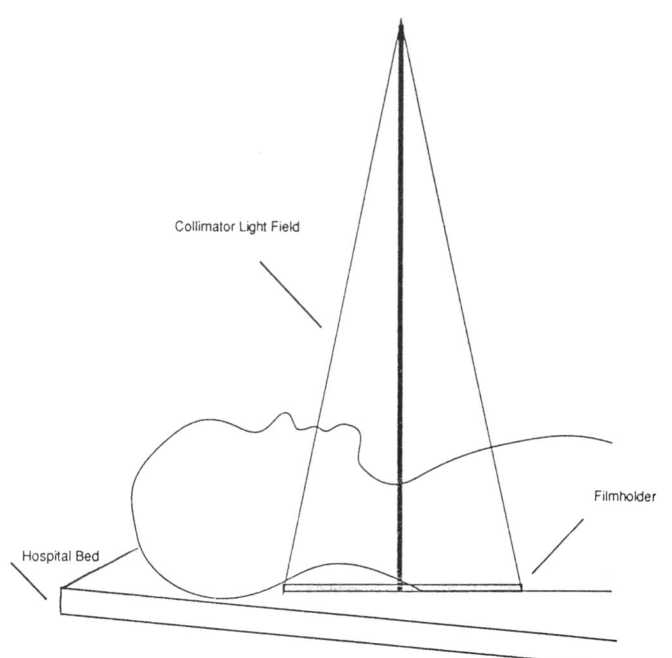

Painted markings denoting the centers of each of the four borders on filmholders are also of great assistance to radiographers in portable radiographic situations (Figure 11-4). This is of special help if part of the film holder is lost from view under the patient (Figure 11-5). Estimates made to determine the center of the filmholder without any kind of visual clue can be off-centered by as much as several inches if the radiographer has limited experience in portable work.

Distortion

Another type of patient misalignment can result even when proper centering is realized. This type of misalignment identifies the subject of "shape" distortion. Shape distortion is the enlargement of different parts of a structure by different amounts. Cahoon (1974) refers to shape distortion as deviations from the true shape of the object as exhibited in its shadow image.

Body parts are three-dimensional objects. Certain anatomical structures are superimposed directly under or over other anatomical structures. The view of some structures requires that the body part be obliqued (rotated or turned) to free that part from the superimposition of the other part.

A classic example is the sternum. To place the sternum close to the film to minimize magnification, the patient is placed in a prone (face-down) position on the table. When radiographed in this manner, the vertebral column is directly over the top of the sternum. By obliquing the patient a few degrees (the amount depending on body size), the sternum is freed from superimposition by the vertebral column.

In many views of complicated skull anatomy a combination of both patient obliquity and tube angulation is used to properly demonstrate the anatomical area of interest, such as the occipital bone of the skull (Figure 11-6). These precise angulations become requirements to satisfactorily demonstrate that portion of the patient's anatomy as requested by the radiologist. The radiographer should note when using an angled central ray that the tube must be angled before the body part or film can be aligned to it. This is an example of how the sequence of logical steps may vary with the radiographic position.

An understanding of the role that angulation plays in shape distortion may help to explain its effect on radiographs. Normally because of object-to-film distance, it is expected that the image will be slightly larger than the object. With angulation of the body part, however, it is possible that the image will be elongated or foreshortened, depending on its alignment. *Elongation* is a lengthening of the image size, whereas *foreshortening* is the amount of the image size reduction. Both are demonstrated in Figure 11-7. An edge view of a disk-shaped object is represented by the line XY. If line XY is parallel to the film, the resulting image will be slightly longer than XY. If XY is angled, the image of XY can actually be shorter than XY, depending on the tilt and its placement under the x-ray tube. Figure 11-7 illustrates the effect of foreshortening. The radiographic effect of foreshortening is demonstrated in the skull radiographs in Figure 11-8. Notice in the right view that certain aspects of the skull appear to be enlarged. This is the effect of a greater object-to-film distance for that part of the skull that is farthest from the film. If the body part is not properly centered under the x-ray beam, the amount of distortion will be altered by the object's distance from the central x-ray beam. Figure 11-9 shows that this

Figure 11-6. Intentional patient rotation and tube angulation. Due to the anatomical complexity of the body many structures are superimposed over other structures. Through the intentional use of tube and patient angulation, structures can be projected clear of others for the purpose of an accurate diagnosis. In the views below, notice the cranium's occipital bone with and without superimposition.

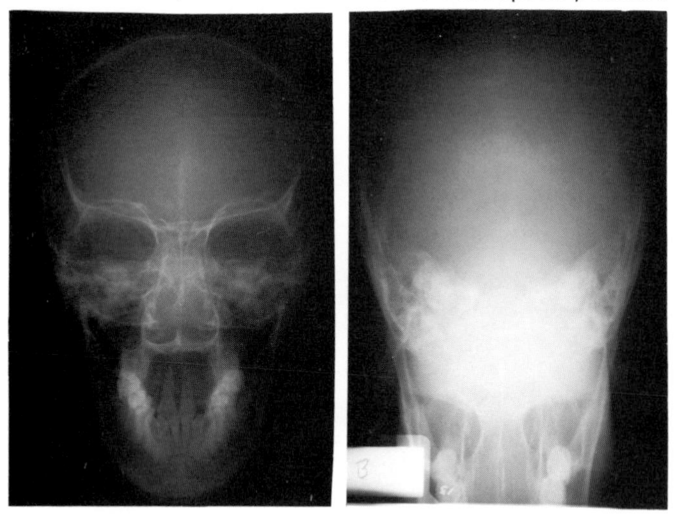

Figure 11-7. Foreshortening effect (side view) of an object denoted by line XY. When parallel to the film, image size is uniform. When angled, image is said to be *foreshortened* (length is reduced) although certain aspects will be greatly magnified, due to object-to-film distance.

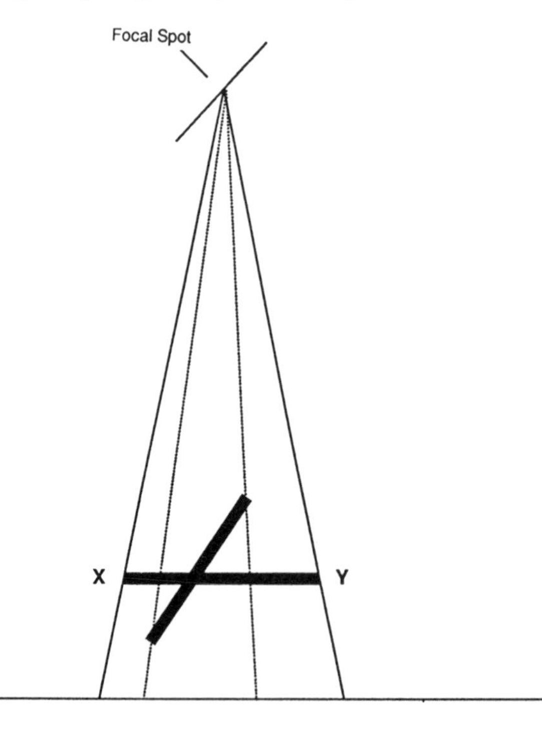

situation can cause either foreshortening or elongation.

Additionally, patient weight or the bed itself may cause the plane of the film to be unparallel to the floor. If this is the case, adjustment of the tube will be necessary to obtain a perpendicular angle to the plane of the filmholder. (There are instances where a perpendicular beam is not the angulation of choice, but is used in the previous statement, because it is the norm in the majority of views done in portable radiography.) Failure to notice that the film (and patient) are angled slightly and *not* adjusting the *tube* angulation to compensate will result in unwanted shape distortion of the radiographic image.

Figure 11-8. Radiographic effect of foreshortening.

With the skull in a "face-down" position, resulting image is shown.

Raising the forehead while lowering the chin results in a foreshortening effect evident by comparing the overall length of the two skull images. (Some magnification of certain anatomical structures occurs due to an increase in object-to-film distance for those structures.)

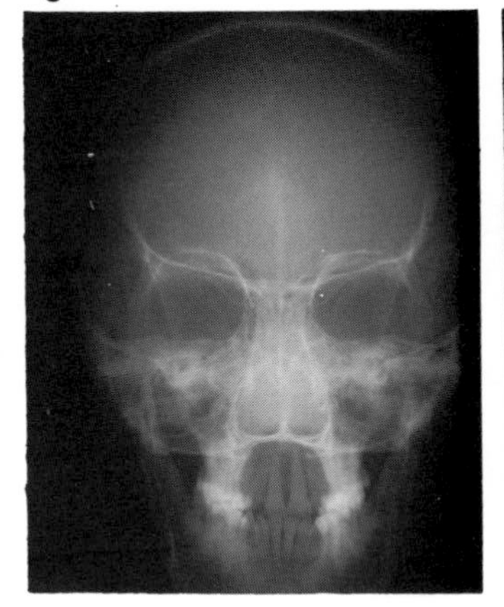

Figure 11-9. Foreshortening compounded by beam angulation. The identical structure projected to film from 3 different angulations. On the left, foreshortening is compounded. On the right, image is elongated.

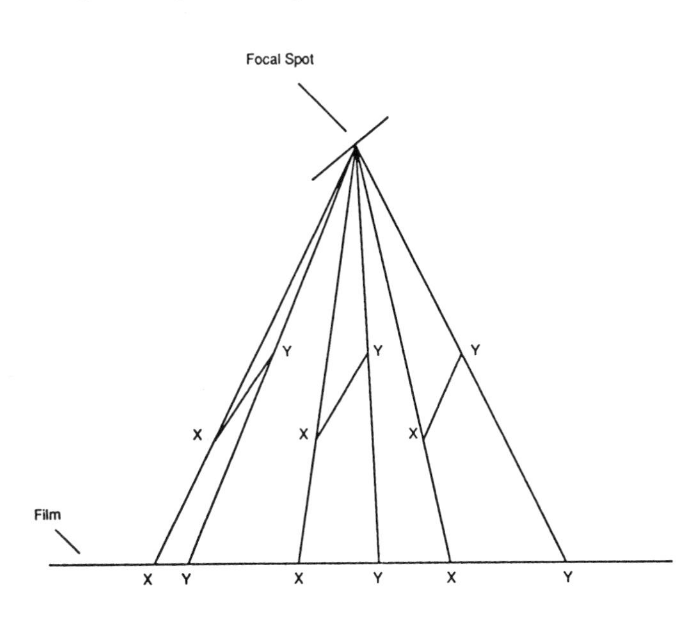

SUMMARY

This chapter has examined the components of alignment: the x-ray tube, body part, and film. All must be geometrically aligned or the result will be a less-than-perfect, if not undiagnostic, radiograph. In tabletop radiography, direct line-of-sight, or visual centering can often be used to ensure proper alignment. Views taken with the filmholder in the Bucky tray can be slightly more complicated, but with a logical sequence of steps and an understanding of the collimator controls and Bucky locks, precise alignment is possible.

Equipment with "floating" tabletops makes mobility of the body part easy and is usually the last step in alignment after the tube and film have been aligned to each other. Open and unsecured Bucky trays were pointed out as avoidable errors. In portable radiography, marked filmholders and framing were described to assist the student radiographer. Framing is the process of centering the film or body part within the borders of the lightfield. Marking the centers of the cassette's borders can be a useful method of centering the filmholder if the edges are visible beneath the patient. Shape distortion was defined at the enlargement of different parts of a structure by different amounts. Due to the requirements of certain views a certain amount of shape distortion may be normal, but dramatic elongation or foreshortening can be the result of misalignment.

STUDY QUESTIONS

1. What is the role of tube-object film alignment in radiography?

2. Describe the alignment problems inherent in portable radiography.

3. Describe how the portable radiographer can insure that the tube is properly aligned to the film in the event that the film is partially hidden from view under the patient. Describe the "framing" technique.

4. If the patient is obliqued 10 degrees off the film for a chest radiograph, describe the difference that will be evident in the two lung fields.

5. Define shape distortion.

6. Using perpendicular x-ray beams and prone or supine patient positions would minimize shape distortion. Explain why this is not always possible for every radiographic view.

7. Describe the sequence of steps to be used positioning with an angled ray.

8. Describe two mistakes that are possible with Bucky work that would result in misalignment, and the solutions for each.

9. Describe the procedure for insuring longitudinal alignment of the central ray of the x-ray tube to the film when the film holder is in the Bucky tray.

10. Explain foreshortening.

BIBLIOGRAPHY

1. Ballinger, Philip W. *Merrill's Atlas of Radiographic Positions and Radiologic Procedures*, 5th ed. St. Louis: C. V. Mosby Company, 1982.
2. Bushong, Stewart C. *Radiologic Science for Technologists, 2nd ed. St. Louis, C. V. Mosby Company, 1980.*
3. **Cahoon,** John B. *Formulating X-Ray Techniques*, 8th ed. St. Louis, C. V. Mosby Company, 1980.
4. *The Fundamentals of Radiography*. Rochester: Eastman Kodak Company, 1980

Chapter 12

The Effects of the Patient on Radiographic Quality

Chapter Outline

1. Learning Objectives
2. Introduction
3. The Role of Tissue Quality on Radiographic Quality
4. The Effects of Tissue Quality on Image Visibility
   A. Radiographic Density
   B. Radiographic Contrast
5. The Effects of Tissue Quality on Image Formation
6. Summary
7. Study Questions
8. Bibliography

Learning Objectives

Upon completion of this chapter, the student should be able to:

1. List the four different body types.
2. Match a given list of characteristics with the appropriate body types.
3. List the factors that influence tissue thickness.
4. List the factors that influence tissue opacity.
5. Explain the relationship between tissue quality and density on the radiograph.
6. Explain the relationship between tissue quality and subject contrast.
7. Explain where and why a contrast medium would be used in the medical field of radiology.

Introduction

Successful production of a radiograph requires the application of a variety of factors including proper machine factors, film screen combinations, processing, and positioning, to name a few. One often underestimated factor that must be considered is the patient. Patients come in different sizes, shapes, and physiques. The composition of the patient requires the radiographer to determine the quantity and quality of radiation required, as governed by the machine factors selected to produce a radiograph of diagnostic quality.

The Role of Tissue Quality on Radiographic Quality

The human body is comprised of different types of tissues. These different types of tissues and their thickness will influence radiation absorption. Radiation absorption corresponds to the atomic number of an element being radiographed. The higher the atomic number, the more dense the element, thus there will be more absorpton of the radiation. Bone for example, has a higher atomic number than muscle making bone less radiolucent.

The physical state of the patient must be evaluated if a good quality radiograph is to be produced; therefore, radiographers need to cultivate the habit of evaluating patients according to their physical and physiological composition. Proper evaluation of a patient should include (but not be limited to) body habitus, type of tissue, pathology, sex, and age.

Generally speaking, human beings can be classified into four groups according to their physique or habitus. These four classifications are hypersthenic, sthenic, asthenic, and hyposthenic. The hypersthenic type is characterized by softness and roundness throughout the various bodily regions. The tissues tend to be flabby with an excess of subcutaneous fat. These individuals have small bones, great strength and a barrel-like torso. The chest is usually well developed, but short and wide at the base and the digestive tract is dominant. Hypersthenic types also have a high waistline, small hands and feet, and a short thick neck, tend to become obese, and have higher rates of hypertension. Other diseases common to this body type include kidney and heart disease.

The sthenic type is strong and active, but movements are relatively slow and awkward because of heavy muscle development. The bones of this body type are large as compared to the hypersthenic build. These individuals usually have a small waist, broad hips, a muscular abdominal wall, and well-developed chest that is larger than their abdomen. In general, persons of this body type tend to be physically and emotionally well balanced.

The asthenic body type is often fragile and delicate. The bones have a tendency to be small, and the facial bones are delicate. The shoulders are usually narrow, and the shoulder girdle lacks muscle support. The scapulae tend to protrude posteriorly. Another characteristic of this body include a flat abdomen with

Figure 12-1. Chart representing the four different habitus.

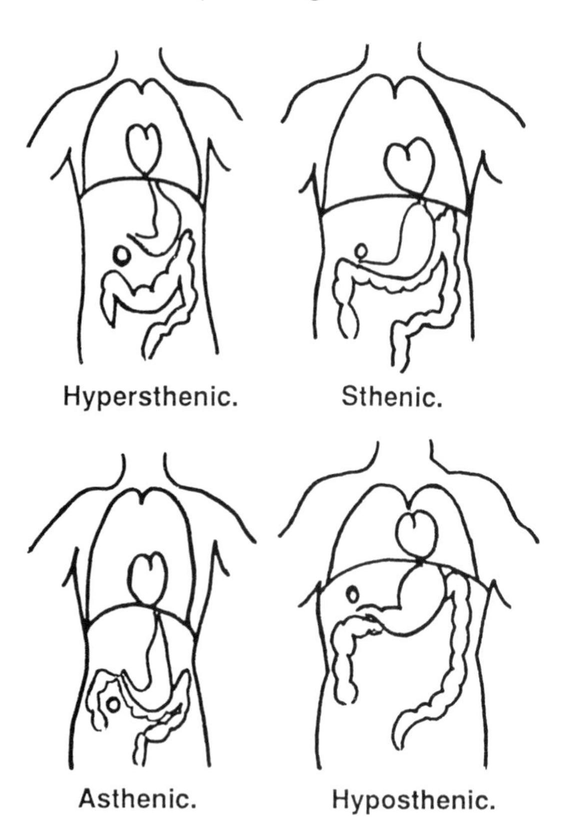

the lower portion of the abdomen being prominent. Usually there is a weakness of the thighs and upper arms. This body type tends to be weak, nervous, and prone to functional nervous disorders.

The last body type is the hyposthenic. Hyposthenic is very similar to the asthenic types, but the stomach, intestines, and gallbladder are situated higher in the abdomen. Figure 12-1 illustrates these four different body types. As discussed above, the position of the body organs, especially the gallbladder, stomach, and colon are located in different places in the four body types. A radiographer obviously needs to know the different body types to be better able to produce a good diagnostic radiograph. It should also be pointed out that individuals may not possess all of the characteristics of a certain body type, but may exhibit several characteristics from several body types. Experience will help determine to which body type an individual should be classified.

Using a broad classification, body tissue can be divided according to organic and organic impregnated with inorganic materials. Organic tissues usually consist of skin, flesh, cartilage, fat and tendons. These tissues are composed of carbon, hydrogen, nitrogen, and oxygen. X-rays can penetrate these tissues very easily. Impregnated organic tissue like bone, dentine, and calcified cartilage are composed of calcium, phosphate, and calcium carbonate. This type of tissue has a high absorption rate making them less penetrable. However, composition of the tissue (organic or impregnated organic) is not the only factor that affects tissue absorption. The thickness and opacity of the tissue will also affect the penetrability of radiation.

Tissues are not only composed of chemical compounds, b have a different molecular structure. The molecular structure muscle cells and red blood cells are very complex as compar to the molecular structure of lung cells. The more complex t cell structure, the more radiation they absorb. The combinati of chemical compound and molecular structure will determi the opacity of the tissue.

Tissue opacity (density) is also influenced by how the tiss is structured and the type of material surrounding the cells. Tiss structure relates to how closely the cells are placed together with the tissue. Another way of saying how close the cells are to ea other would be "mass per unit volume" or "compactness" the cells. Muscle cells are very compacted and complex in the structure. The red blood cells are complex in their structure, b the cells are not close together. Comparing these two cells a cording to complexity and compactness, muscle is said to denser than red blood cells. In comparison, bone cells have complex structure, but a lower mass per unit volume than mu cle. Why is it, then, that bone absorbs more radiation than mu cle? The reason for this is that between bone cells there is materi consisting of chemical compound that fills in the vacant space This makes the bone more dense than muscle.

Tissue opacity can be influenced by the status of organs th are to be radiographed. Organs are classified as being hollow like the stomach, or solid, like the liver. Hollow organs are hard to distinguish on radiographs because there is little tissue vari tion between the organ and surrounding tissues. The radiogra will exhibit the same density for each because they are almo identical. If a hollow tissue is filled with a gas, liquid, or soli the opacity will change and the organ will be better distinguish on the radiograph. One of the ways to demonstrate hollow orga on a radiograph is to fill these organs with a contrast mediun

Contrast media which increase the photon absorption are terme radiopaque (positive), and are made from substances with a hig atomic number such as, barium sulfate or iodine. Other contra media which decrease photon absorption are known as radiol cent (negative), and are substances with a low atomic numb like different types of gases.

Figure 12-2 illustrates this concept with three radiographs the colon. The colon is a hollow organ and can best b demonstrated radiographically with the use of a contrast mediun Radiograph A is an anterior-posterior view of the abdomen, which the colon is not demonstrated. Shadows on the radiograp may allow partial visualization of the colon. In Radiograph B radiopaque/positive contrast medium (barium sulfate) was use to fill the colon. Notice that the colon is demonstrated and ap pears as a light/white radiographic density. The radiologist ca now determine through better visualization of the imaged detai if there are any abnormalities within the colon. Radiograph demonstrates the presence of gas in the colon, which is a radiolu cent contrast medium. In this case, the use of a lower atomi number contrast medium causes the colon to appear as darker/black radiographic density.

The thickness of the tissue is the other factor that will affe

Figure 12-2.

FIGURE 12-2A.

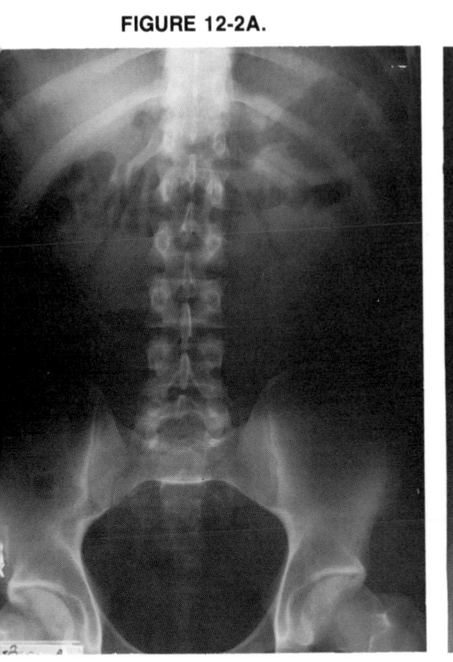

P of abdomen— colon is not emonstrated.

FIGURE 12-2B.

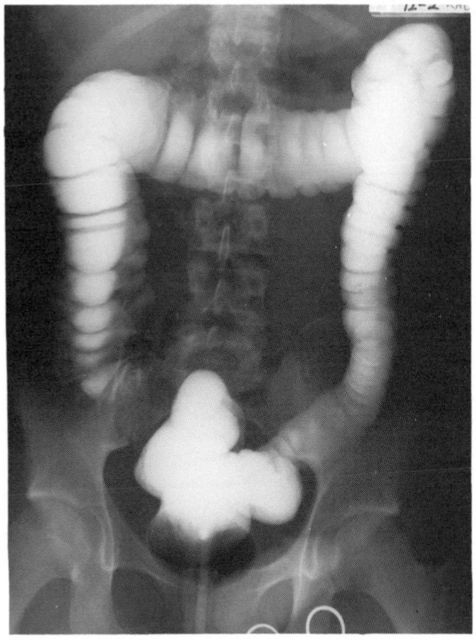

Contrast medium in colon, Fully demonstrated.

FIGURE 12-2C.

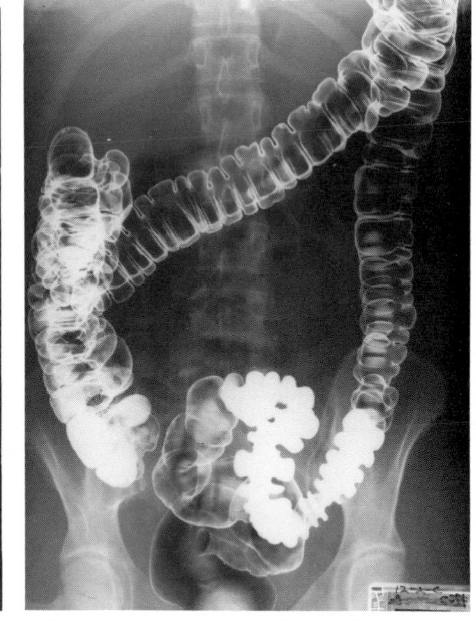

Gas is demonstrated in colon.

ie penetrability of the tissue. Thickness can be influenced by ge, sex, body type, and pathology. For example, radiographing child's knee which measures 6-7 centimeters compared to an dult's knee measuring 12-13 centimeters requires a change in e quantity of x-rays to produce an adequate radiograph. Both hild and adult knees have the same type of tissue; however, the dult has a thicker part which requires an increase in the x-ray xposure. Figure 12-3 shows two radiographs of an anterior-osterior view of the knee. Radiograph A demonstrates the adult's nee. The factors of 74 kVp, 6 mAs, 40″ FFD were used. adiograph B demonstrates the child's knee using the same chnical factors as for the adult's knee. The difference in the mount of density on the film is due to the thickness difference f the two knees.

Another factor that radiographers should consider when dealg with thickness of the body part is the sex of the patient. emales in general have more subcutaneous fat in the pelvic, uper femoral, and arm regions than males. Conversely, males tend be more muscular and their boney structure is more massive an the females. Consequently, when radiographing female and ale patients the radiographer should take these differences into onsideration when determining the technical factors. Radioraphing a male's shoulder, for example, would require more chnical factors than a female's because the bony structure and uscles are more massive or more dense.

The pathological status of the body part is yet another influncing factor on thickness. Since thickness relates to the volume f tissue, some diseases can actually increase or decrease the hickness of the part being radiographed. The physiological conition of the patient is probably the hardest factor to predict when aking the proper radiographic technique allowances. The best way to acquire knowledge of a patient's condition is to read the request slip and attempt to determine why the ordering physician wants the radiograph. The patient is sometimes a good historian, but may not be aware of why the radiographs are being produced. Table 12-1 illustrates the different types of pathology with a guide to either increase or decrease the machine factors. This chart may prove useful when trying to determine exposure adjustments based on pathologic conditions of patients.

Figure 12-3.

FIGURE 12-3A. **FIGURE 12-3B.**

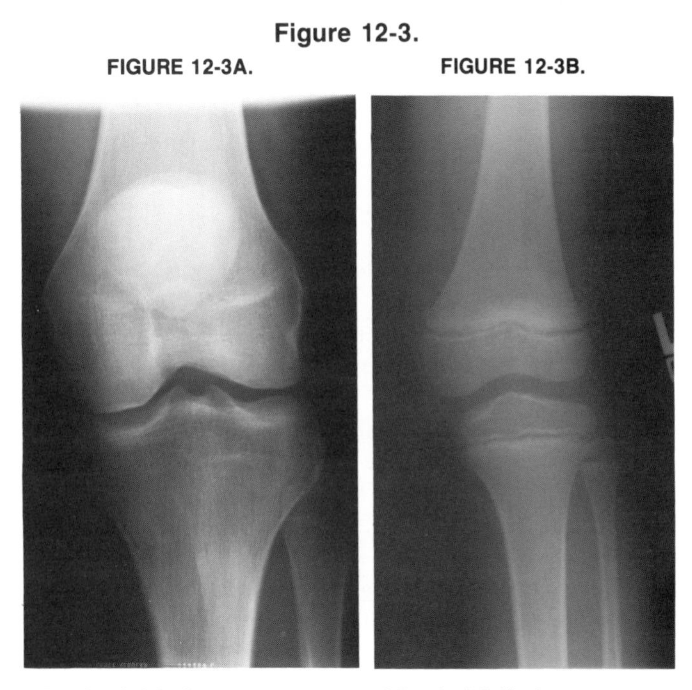

AP of adult's knee. AP of child's knee.

TABLE 12-1.

SYSTEM	PATHOLOGY	
	ADDITIVE	DESTRUCTIVE
Skeletal System	Acrogmegaly Kyphosis Charcot joint Exostosis Hydrocephalus Marble-bone Metastasis Osteochondroma Osteoma Padet's disease Sclerosis	Osteomyelitis Tuberculosis Nercrosis Atrophy Blastomycosis Carcinoma Coccidionmycosis Cystic condition Arthritis Ewing's tumor Hodgkin's disease
Respiratory System	Actinomycosis Atelectasis Edema Malignancy	Lung abscess Emphysema Pneumothorax Tuberculosis
Circulatory System	Ascites Cirrhosis	Gout Leukemia
Gastrointestinal System		Bowel Obstruction Malignancy
Soft Tissue	Edema	Emaciation Pneumoperi- toneum

ADDITIVE—Hard to penetrate—increase technical factors
DESTRUCTIVE—Easy to penetrate—decrease technical factors

The Effects of Tissue Quality on Image Visibility

Radiographic Density

Tissue quality is a major factor affecting radiographic density. When talking about tissue quality, one has to consider the atomic number of the elements that make up the tissues, complexity and compactness of cells, if the tissue is diseased, and where in the body the tissue is situated.

Stop and think about the different types of cells and tissues located in the chest. The chest, although large, is not as large as some parts of the body and is relatively thin, for example, compared to the thickness of the abdomen. The chest is composed of soft tissue, ribs, thoracic spine, clavicles, heart, and lungs. Each one of these parts has a different tissue quality, although some may have a tissue quality that is similar to others. With each different tissue there will be a change in the amount of density on the radiograph. Figure 12-4 demonstrates a posterior-anterior view of the chest. Compare the different levels of density with regard to the lungs, ribs, and thoracic spine. The lungs, because of the type of cells and tissues that make up their composition, are considered to be easily penetrated. That is, the radiation penetrates the lungs easier than it would a more dense tissue like the thoracic spine. The conclusion can be made that the more opaque the tissue, the less dense its appearance on the radiograph. Consequently, there is an inverse relationship between the opacity of the part being radiographed and density on the radiograph.

Most radiographers have found that the chest has unique problems associated with the different body habitus. Therefore, radiographers should be aware that because of the different body habitus, there are several types of thoracic cavities. Each of these thoracic cavities will require a change in the exposure factors for the desired amount of radiographic density on the radiograph.

The first thoracic cavity is classified as a *heavy chest*. Usually the patient has a very muscular frame and is very thick. Women who have a large frame and large breasts would fit into this category. This type will require an increase of the exposure factors. The next type is classified as a *thin chest*. This patient has a lack of flesh on the ribs, and the scapulae protrudes when positioned. This type of patient would require a decrease of exposure factors. The last type is classified as a *very thin* type. This patient has a barrel-like chest and the thoracic cavity represents the entire depth of the thorax. This type patient will also require a decrease of the exposure factors. Remember, these types of thoracic cavities are different from the normal thoracic cavity.

Referring to the chart in Table 12-1 under the respiratory system, malignancy requires an increase of the exposure factors. Malignancy is an abnormal growth of tissue. This growth can occur in any part of the body. Since this growth is a "growing" tissue, it will make the part being radiographed more dense (opacity). Figure 12-5 shows two radiographs of a posterior-anterior view of the chest. Radiograph A demonstrates a chest with a malignancy. The area of the malignancy is very light and lacks radiographic density. Radiograph B demonstrates a chest that is considered to be normal. There is a uniform density throughout the chest.

Both patients were of the same body type, thickness, and sex. The only difference between the two patients is one has a malignancy, the other does not. If the technical factors had not been increased in Radiograph A, there would not have been the proper amount of density on the radiograph to demonstrate the

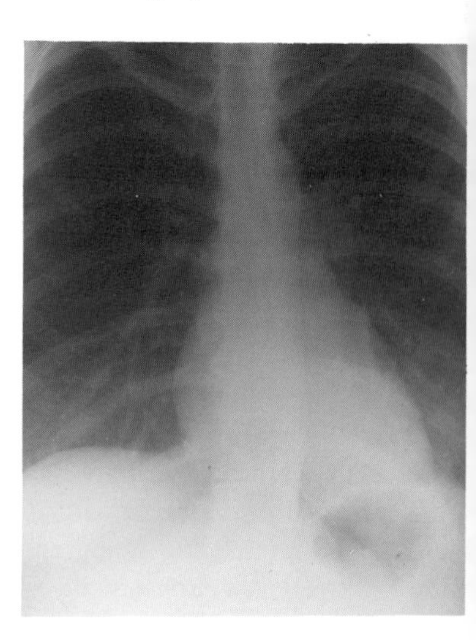

Figure 12-4. PA Chest Radiograph. Notice the different amounts of density in relationship to the lungs, ribs, and spine. The denser the tissue the less amount of density on the radiograph.

Figure 12-5.

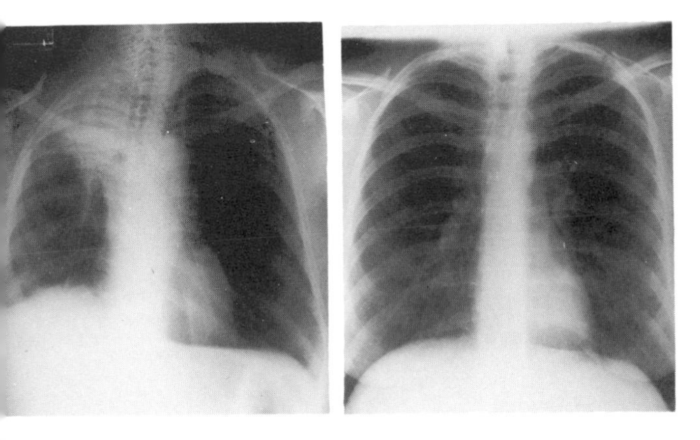

PA chest with malignancy. There is less density in the area of the malignancy.

PA chest with no malignancy. There is uniform density throughout the chest.

malignancy sufficiently. One should be aware that if the malignancy occurs in other parts of the body, such as the bones, this abnormal growth destroys the bones and the technical factors will have to be decreased.

Radiographic Contrast

Subject contrast is defined as the differences in the quantity of radiation transmitted by a particular body part as a result of the different absorption characteristics of the tissues and structures making up that part (Donohue, 1980). The type of subject contrast will vary given various body parts. In some areas of the body there will be greater differences of radiographic density between the different tissues. This is considered *high subject contrast*. In other areas, the tissue is of the same thickness and opacity producing a film with very few changes in the radiographic density. This is considered *low subject contrast*.

Figure 12-6 shows three radiographs with different subject contrasts. Radiograph A is an anterior-posterior view of an abdomen. The tissues of the organs located in the abdomen are basically the same opacity and within the abdomen are many organs that the radiographer must demonstrate. There are shadows of some of these organs, such as the stomach, gallbladder, liver, and colon. How then, for example, can the gallbladder be fully demonstrated? The patient is usually administered an oral contrast medium that can be absorbed by the gallbladder. The gallbladder is now very opaque (dense) and absorbs the radiation. Since the contrast medium is more opaque, the gallbladder will appear as a white area on the radiograph when a scout film of the abdomen is done. The areas aaround the gallbladder are not as opaque and the radiation penetrates this area, thereby creating more density on the film. Radiograph B demonstrates a patient's abdomen after being administered an opaque contrast medium. The gallbladder is visualized and there is high subject contrast. Radiograph C is an anterior-posterior view of the shoulder. The tissues that make up the shoulder have different opacities; therefore the amount of density is greatly varied creating a high subject contrast without the use of a contrast medium.

The use of a radiolucent contrast medium such as air can be used to increase subject contrast. Why is a chest radiograph taken on inspiration? How does the radiologist know that the patient did not fully inhale before the radiograph was taken? The lungs are like sacs that inflate and deflate with respiration. One might

Figure 12-6.

FIGURE 12-6A.

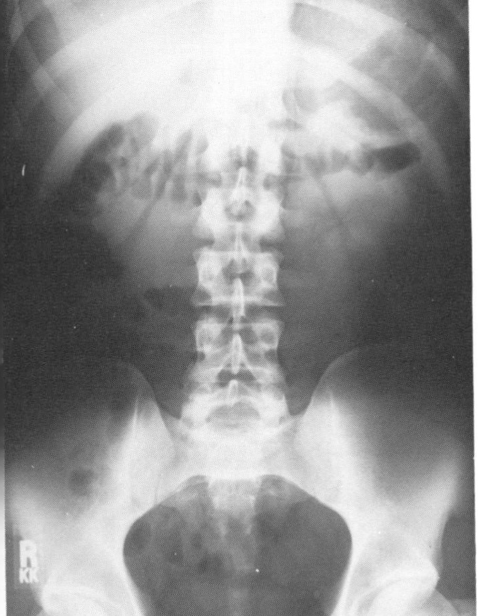

AP abdomen with low subject contrast.

FIGURE 12-6B.

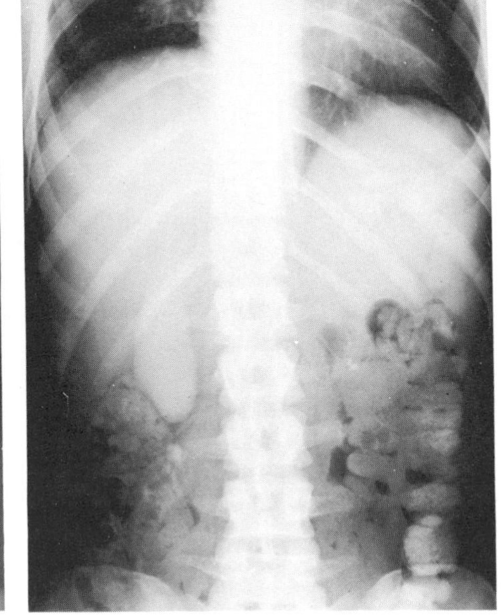

AP abdomen after the administration of a contrast medium. The gallbladder is now visualized.

FIGURE 12-6C.

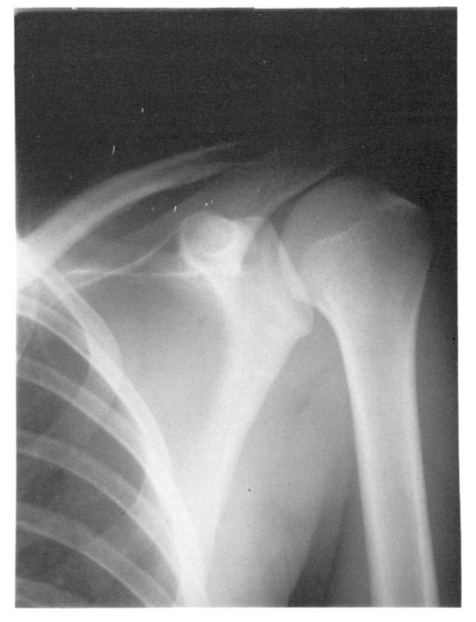

AP shoulder with high subject contrast.

Figure 12-7. Expiration (left) and inspiration (right) of infant chest.

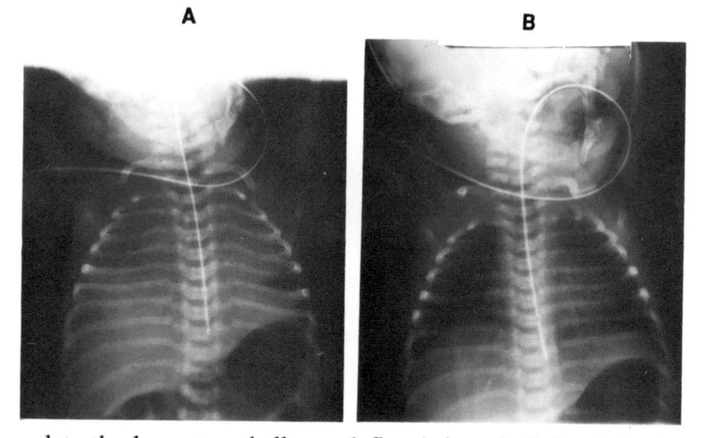

relate the lungs to a balloon: deflated there is little air in the balloon, but after inflating the balloon with air the balloon gets larger. Remember the different tissue quality of the chest, one can relate this to the lungs. If a radiograph is taken on expiration the lungs have little air and the thoracic (chest) cavity has little subject contrast. When the radiograph is taken on full inspiration, the lungs are full of air. Air is very radiolucent; therefore, an area where air is located will have more radiographic density on the image.

But why is it important that the lungs be full of air? Radiologists use the base of the lungs to help determine if an individual has an enlarged heart or other abnormalities. This is the reason that the patient should have full inspiration for a chest radiograph. Figure 12-7 demonstrates this concept. Radiograph A is an anterior-posterior of the chest on an infant. The radiograph was taken on expiration. Radiograph B on the same infant was taken on inspiration. In radiograph A the lung fields cannot be seen because there is no air in the lungs. Since the surrounding areas of the lungs are the same tissue quality, the lungs are not demonstrated. On Radiograph B taken on inspiration the lungs are full of air. Since air is radiolucent, the lungs are demonstrated on the radiograph.

The Effects of Tissue Quality on Image Formation

Generally speaking, the quality of patient tissue does not affect recorded detail, size or shape distortion.

SUMMARY

Reviewing Table 12-2 one can see that tissue quality affects the amount of density on the radiograph. Increasing the part thickness or tissue opacity, decreases the amount of density on the radiograph; decreasing the part thickness increases the amount of density on the radiograph (inverse relationship). Tissue quality also affects radiographic contrast. As tissue quality increases, radiographic contrast increases; as tissue quality decreases, the subject contrast decreases (direct relationship). When radiographing parts of the body that have the same tissue opacity, subject contrast can be enhanced by the use of a contrast medium. Remember, contrast media are classified as either positive (radiopaque) or negative (radiolucent). Tissue quality has no significant effects on the remaining factors of recorded detail or distortion. To ensure a satisfactory diagnostic radiograph the first time, radiographers need to remember to evaluate their patients in regards to body habitus, sex, tissue opacity, and pathology.

TABLE 12-2. **Summary of the Effects of Tissue Quality on Radiographic Quality.**

IMAGE VISIBILITY	
Radiographic Density	–
Radiographic Contrast	+
IMAGE FORM	
Recorded Detail	0
Size Distortion	0
Shape Distortion	0

+ = Direct Relationship to Image Characteristic
– = Inverse Relationship to Image Characteristic
0 = No Effect on Image Characteristic

STUDY QUESTIONS

1. List the four different body builds or habitus.

2. Match the characteristic in Column B with the body type in Column A.

A	B
Hypersthenic	A. Bones are large and have physique
Asthenic	B. Have small hands and feet. Neck thick and short.
Hyposthenic	C. Fragile and delicate with narrow shoulders.
Sthenic	

3. List three factors that affect tissue thickness.

4. List three factors that affect tissue quality.

5. Using the words increase, decrease, remains the same, explain what happens to density, contrast, recorded detail, and distortion when the tissue quality is increased.

6. Name two contrast media that are used in radiography and state if they are positive or negative contrast.

BIBLIOGRAPHY

1. Cahoon, John B. *Formulating X-Ray Techniques* 6th ed. Durham, N.C.: Duke University Press, 1974.
2. Donohue, Daniel P. *An Analysis of Radiographic Quality*. Baltimore, Maryland: University Press, 1981.
3. Fuchs, Arthur W. *Principle of Radiographic Exposure & Processing*. Springfield, Ill.: Charles C. Thomas, 1973.
4. Jacobi, Charles A. *Radiologic Technology*. St. Louis: C. V. Mosby Co., 1968.
5. *The Fundamentals of Radiography*. Rochester: Eastman Kodak Company, 1980.

Chapter 13

The Effects of Grids on Radiographic Quality

Chapter Outline

1. Learning Objectives
2. Introduction
3. Grid Construction
4. Terminology
5. Elimination of Grid Lines
6. Grid Cutoff
7. The Effect of Grids on Image Visibility
   A. Radiographic Density
   B. Radiographic Contrast
8. The Effect of Grids on Image Formation
   A. Recorded Detail
   B. Size and Shape Distortion
9. Summary
10. Study Questions
11. Bibliography

Learning Objectives

Upon completion of this chapter, the student should be able to:

1. Describe the role of grids in radiography.
2. Compare parallel, focused, and cross-hatch grids.
3. Identify definitions of grid ratio, grid focus, and grid radius.
4. Compare wafer grids and grid cassettes.
5. Identify the method of eliminating grid lines form appearing on a radiograph.
6. Identify possible causes of grid cutoff.
7. Given a change in grid ratio, identify its effect, if any, on:
   a. Radiographic density.
   b. Radiographic contrst.
   c. Scale of contrast.
   d. Recorded detail.
   e. Size distortion.
   f. Shape distortion.
8. Given different grid ratios, identify the one that would produce a radiograph with more (or less) density, higher (or lower) contrast, longer (or shorter) scale of contrast, as specified.

Introduction

As anatomical tissue absorbs x-radiation, complicated physical interactions take place wherein new x-rays are totally absorbed by tissue or are modified. Both secondary and scatter x-radiation account for the large amount of "fog" on the image. Fog is defined as additional density on a radiographic image usually caused by secondary or scatter radiation produced within the body part. This can be substantial enough for some larger or thicker body parts to be detrimental to the quality of the diagnostic image. Methods have been mentioned which are of critical importance in limiting the amount of primary radiation that the patient receives, but the most effective way to remove the scatter radiation is with a device called a grid (Figure 13-1).

Grid Construction

In 1911, Dr. Gustave Bucky built the first grid using strips

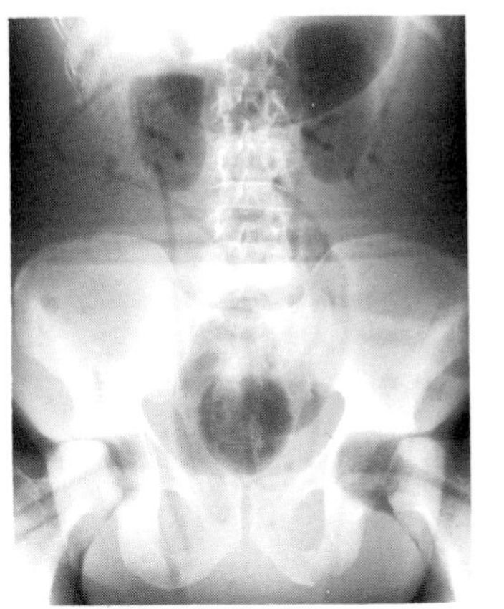

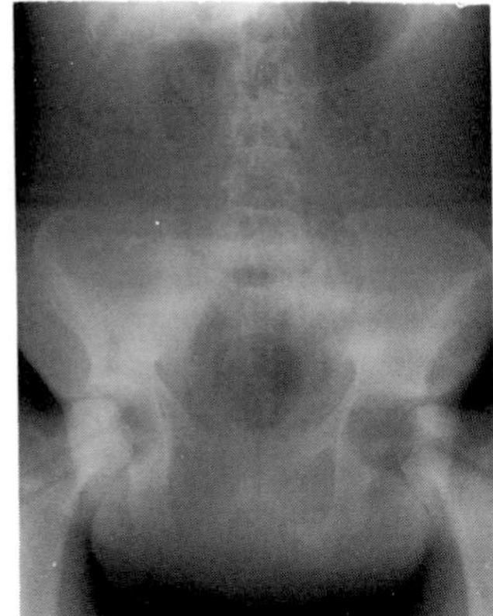

Figure 13-1. At the left is an abdomen radiograph taken with the use of a grid. On the right is an abdomen radiograph with equivalent density, but taken without the grid. Notice the extreme amount of fog (grayness) caused by excess scatter and secondary radiation reaching the film.

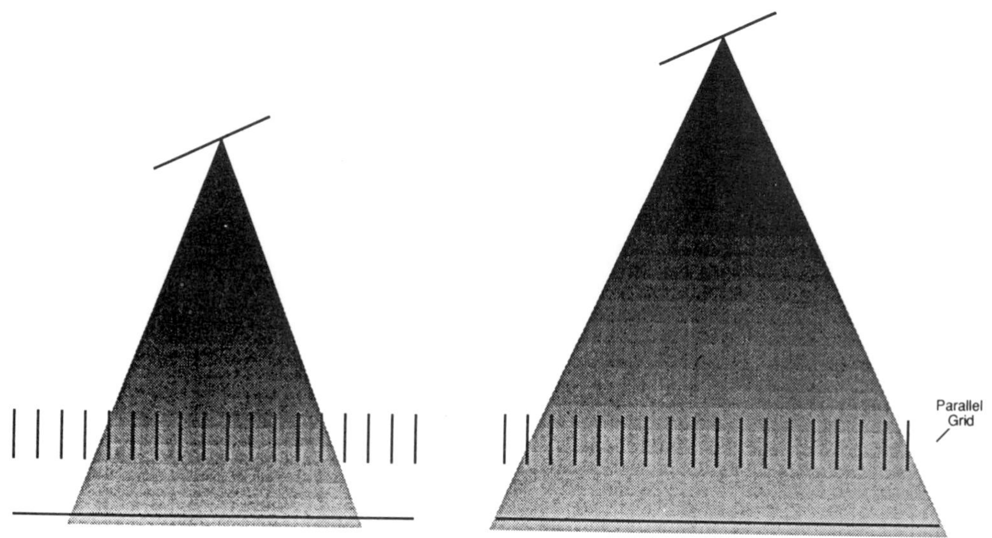

Figure 13-2. Unfocused or Parallel Grid. Since the lead strips are not angled, infinitely long target-to-film distances can be used with a parallel grid, as illustrated on the right. Their use in modern clinics is limited, however, because with short TFD's the grid will eliminate almost all primary radiation at the side edges of the radiograph.

of lead foil standing on edge separated by radioparent interspacers (Liebel-Flarscheim). While the basic principle remains unchanged there have been substantial improvements in design, quality, and ease of use of grids throughout the years.

The simplest type of grid is called a parallel, or unfocused, grid. The strips in a parallel grid, as the name implies are all parallel to each other. The only strips that coincide exactly with the direction of the primary x-rays are those exactly under the central ray, which is the greatest disadvantage of this type of grid. As may be seen in the accompanying Figure 13-2 the lead strips closer to the edges make shadows wider than themselves. Only directly under the tube does the width of the shadow match the lead width. This cutoff of primary radiation results in lighter densities at the edges of the film. Uniform density films cannot be obtained using parallel grids if the film is more than 6 or 8 inches wide at normal FFD's. Parallel grids are only appropriate, therefore, with small film sizes or long target-to-film distances (according to the Liebel-Flarscheim Company). Because of this, parallel grids are seldom encountered in modern radiographic environments. (Discussions to follow will not be dealing with parallel grids unless specifically mentioned.)

Most grids today consist of hundreds of lead strips carefully positioned, *angled*, and separated. These grids are referred to as focused grids because each grid strip is uniquely angled to coincide with the direction of x-rays being emitted from the x-ray tube (Figure 13-3). This perfect alignment allows most primary rays to pass between the lead strips (inevitably, some will collide with the top edges of the strips). The interspace material for spacing the lead strips should ideally stop no x-radiation. Since there is no such material, minimal absorbers of low density are used, such as fibre, plastic, or aluminum. Most of the scatter radiation entering the grid from different angles is likely to strike one of the lead strips, and therefore *not* get to the film. Both focused and unfocused grids can be referred to as linear grids.

A crossed, criss-cross, or cross-hatch grid is actually two linear grids, focused or unfocused, one on top of the other, with the lead strips of the top grid running at right angles to the strips of the bottom grid (Figure 13-4). Although the secondary and scatter absorption of cross-hatch grids is good, these grids are

Figure 13-3. Grid is used with body parts that produce large quantities of scatter radiation. Note that grid absorbs scatter that would otherwise reach and fog the radiograph.

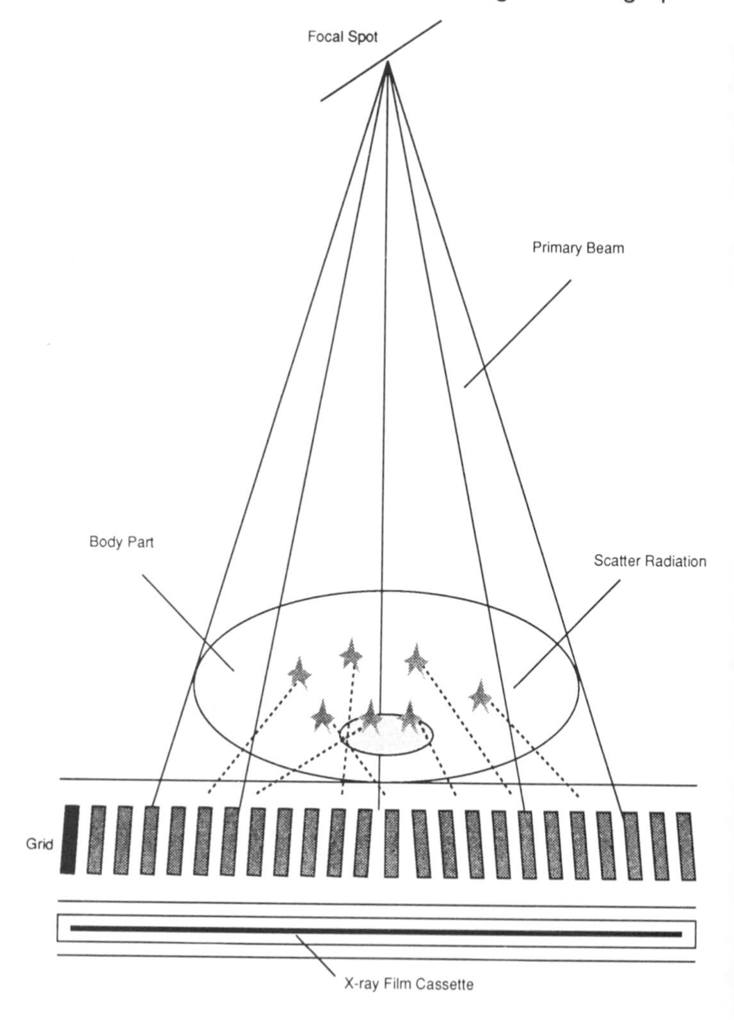

Figure 13-4. Comparison of linear grid to cross-hatch grid. Linear grid contains one set of grid strips (focused or parallel) all running in the same direction. Cross-hatch grid contains two grids, one on top of the other, at right angles to one another.

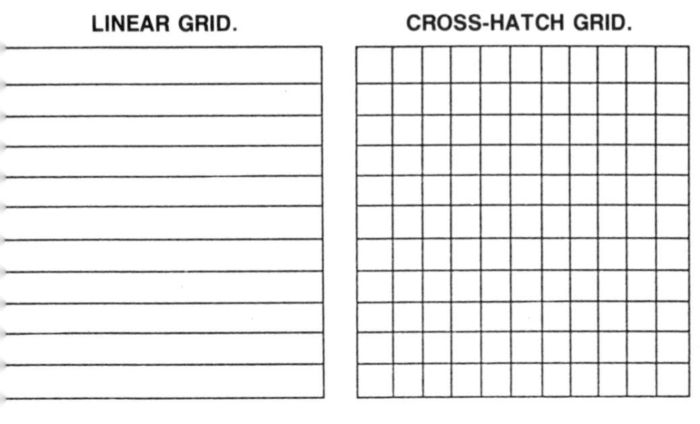

not widely used in conventional diagnostic radiography because of a limitation that shall be described.

Terminology

The number, height, thickness, and proximity of lead strips are all factors in scatter radiation elimination. The *ratio* of a grid is the comparison of the height of the lead strips to the distance between the lead strips. One indication of a grid's efficiency is its ratio. If a grid contains strips that are 60 thousandths of an inch in height and 5 thousandths of an inch apart, the ratio of this grid is 12:1 (Figure 13-5).

Grids are manufactured in different ratios ranging from 5:1 to 16:1. Scatter radiation is absorbed to varying degrees by grids depending on the ratios. Grids' absorption efficiencies expressed as a percent are approximately as follows: 5:1, 75 percent; 6:1, 85 percent; 8:1, 90 perent; 12:1, 95 percent; and 16:1, 96 percent (Cahoon, 1974). Although grid efficiency increases with higher grid ratios, the resultant film is also made lighter because larger quantities of scatter radiation are being absorbed. This will require increased exposure to maintain adequate radiographic densities. Higher grid ratios, although more efficient, are not always the preferred type since other factors in the selection of a grid need assessment.

Thinness of the lead strips will be a factor in reducing the strips' visibility to the eye. The number of lines per inch will also be an indication of the visibility of the grid lines. As the number of lines per inch increases, absorption of primary radiation increases provided lead thickness remains the same and interspace material is made thinner. The increased radiation absorption is due to the increased amount of lead in the grid. Absorption of primary x-rays may be reduced by making the strips thinner; however, scatter absorption would also be reduced. A 12:1 grid made with thinner lead will have absorption comparable to an 8:1 or 10:1 grid using thicker lead. There is a limit to how thin the lead strips can be. With more than about 120 lines per inch, the lead is so thin that high-intensity scatter radiation passes right through the strips.

Another indication of the efficiency is indicated by the lead content, commonly expressed in gr/cm². This is the weight in grams of lead per square centimeter. The higher the lead content, the greater the grid's ability to remove secondary and scatter radiation. There is a link between lead content, grid ratio, and number of lines per inch. If the grid ratio is unchanged and the lead content is increased, it is because the number of lines per inch has decreased. To keep the ratio unchanged and to have more lead would mean that the strips are taller or thicker. If the strips are taller, the interspace must be increased (to keep the same ratio). If the strips are thicker, the interspace remains the same, but either way, less lines per inch will be the end result.

The grid focus is the point above the grid where the strips of the focused grid converge (Figure 13-6). The grid radius is the distance from the grid focus to the grid (Figure 13-7). If a matching target-to-film distance is used, a minimum of primary radiation will be intercepted. When the distances are mismatched, more primary radiation will strike the lead strips instead of passing through them. Only by checking the grid can a radiographer know what FFD is appropriate for the grid that has been selected. Grids are usually marked by the manufacturer as to ratio, radius, and the number of lines per inch. There is some margin for error in FFD use, but the amount of leeway is relative to the grid ratio. Higher ratio grids allow less margin of error than lower ratio grids (Table 13-1).

The types of grids generally used for portable radiographic examinations are wafer grids and grid front cassettes. A wafer grid

Figure 13-5. This ratio represents the comparison of the height of the lead strips divided by the space between the lead strips. In this example, since the height is .060″ and the interspace is .005″, the grid ratio is:

$$\frac{.060}{.005} = \frac{60}{5} = \frac{12}{1} = 12:1$$

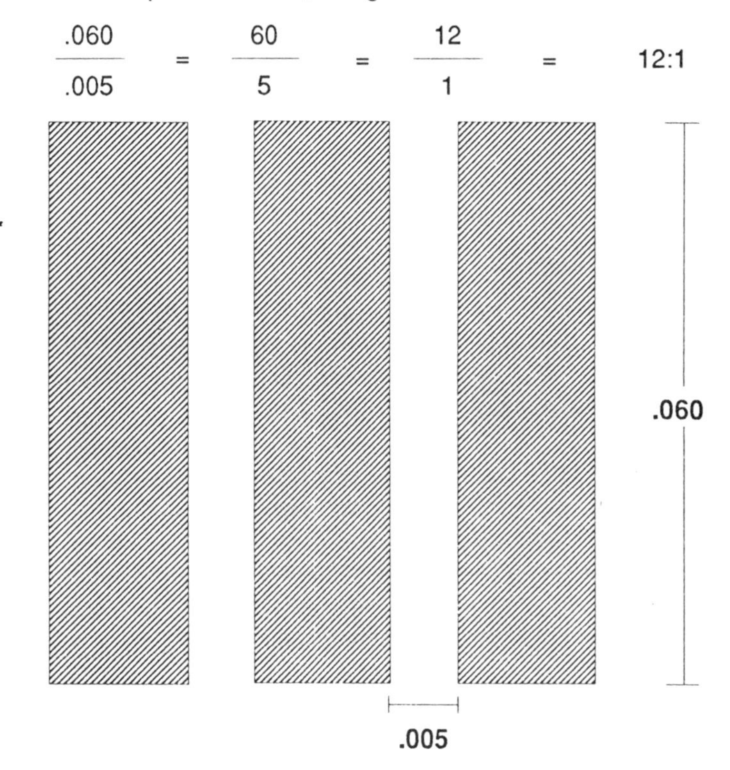

TABLE 13-1. **Summary of Grid Applications.**

GRID APPLICATION GUIDELINES	
4:1 ratio, linear (Low Dose)	Special grid for use in image intensification and spotfilming. Adequate cleanup for small cone-down areas combined with relatively small patient dosage.
5:1 ratio, linear	Moderate cleanup. Extreme latitude in use. Use at lower kilovoltages (up to 80 kVp), wherever wide latitude is desired. Very easy to use.
5:1 ratio, crossed	Very high cleanup, especially at lower kilovoltages. Extreme latitude in use. Use up to 100 kVp, wherever wide latitude and excellent cleanup are desired. Very easy to use. Not recommended for tilted-tube techniques.
6:1 ratio, linear	Cleanup slightly better than the 5:1 linear, but with approximately the same positioning latitude.
6:1 ratio, crossed	Better cleanup than the 5:1 crossed. Very easy to use. Not recommended for tilted-tube techniques.
8:1 ratio, linear	Better cleanup than 6:1 linear. Fair distance latitude. Little centering and leveling latitude. Use up to 100 kVp where wide latitude is not required.
10:1 ratio, linear	Better cleanup than 8:1 linear. Very little positioning latitude. Use for both low and high kilovoltage techniques (up to 110 kVp or slightly higher). Extra care required for proper alignment in use. Usually used in fixed mount or Potter-Bucky Diaphragm.
12:1 ratio, linear	Better cleanup than 8:1 linear. Very little positioning latitude. Use for both low and high kilovoltage techniques (up to 110 kVp or slightly higher). Extra care required for proper alignment in use. Usually used in fixed mount or Potter-Bucky Diaphragm.
16:1 ratio, linear	Very high cleanup. Practically no positioning latitude. Intended primarily for use above 100 kVp in Potter-Bucky Diaphragm. Excellent for high kV radiographs of thick body sections.

is the grid by itself. Manufactured in various ratios with various grid radii and in all sizes (to match film holders), they are used by being taped or secured to the front of a regular film holder. Wafer grids are also available with a bracket or sleeve that allows them to be slid over the top of a conventional film holder. There is a minimum of time spent in attaching them, and they have the advantage of being used for as many exposures as need to be made. The grid front cassette (or grid cassette) is a grid and filmholder combined. This special filmholder has the advantage of already being prepared, and thus, no time is spent in securing the grid. The disadvantage is that before this grid can be used again the cassette must be reloaded with a fresh film.

Elimination of Grid Lines

Grid lines will be seen on radiographs taken with wafer grids and grid cassettes. Both of these grid types are considered to be *stationary* (nonmoving) grids.

Because of the interception of primary rays, it is logical that the appearance of grid lines would make identification of radiographs taken with grids easier (Figure 13-8). The appearance of such lines, however, could obscure or detract from small quantities of diagnostic information. In 1920, Dr. Hollis Potter overcame this problem by producing a device which causes a grid to move at right angles to the direction of the grid strips. This makes the grid lines invisible if correct operating procedures are observed with the Potter-Bucky diaphragm.

Today this device, commonly referred to as the "Bucky," is a sliding tray-like device found just below the table on which a patient lies during the x-ray examination (Figure 13-9). The film is held securely in place by a pair of jaws that center the film under the midline of the table once the tray is pushed firmly to the closed position. The grid moves crosswise from side to side at a rapid speed and is activated by the depression of the exposure switch at the control panel. The movement begins a fraction of a second before x-rays are produced and is terminated as the ex-

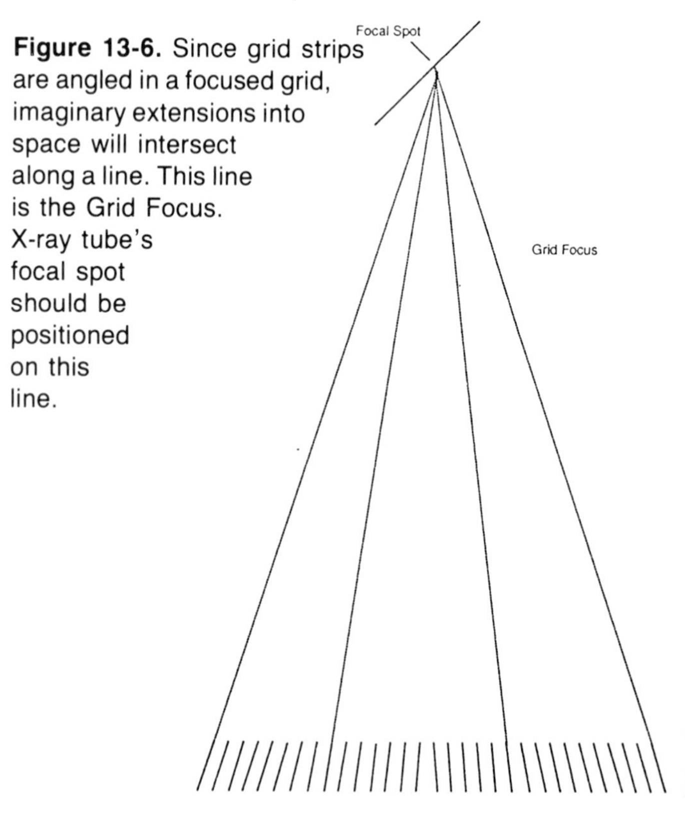

Figure 13-6. Since grid strips are angled in a focused grid, imaginary extensions into space will intersect along a line. This line is the Grid Focus. X-ray tube's focal spot should be positioned on this line.

posure button is released. This ensures movement during the entire exposure (Figure 13-10).

If the Bucky mechanism is working properly, it will make little difference to the radiographer if it is the reciprocating, recipromatic, or oscillating type, but there are minor technical differences. The reciprocating type is as described above, side-to-side movement. This movement is caused in one direction by a solenoid. The return, usually half as fast, is brought by a spring pulling against an oil-filled piston. The movement of a recipromatic grid is also a side-to-side movement, but is propelled by an electric motor. Both strokes are equal in time. Oscillating grids move in a more circular motion as the grid bounces off springs in the corners of its holder.

Grid Cutoff

With stationary and moving grids, angulation of the central ray is possible as long as it is along the direction of the grid strips (Figure 13-11). If using the Bucky tray, this means that the tube may be tilted toward either end of the table as long as it is over the centerline of the grid. Angling "across" the grid strips, which means angling toward the front or rear of the table, would be one of the situations producing grid cutoff. Grid cutoff is a problem where the grid unduly removes primary radiation leaving the radiograph dramatically underexposed at one or both side edges (Figure 13-12). Angling toward the ends of the x-ray table, or the ends of the grid is permitted with all except the cross-hatch grid. With the cross-hatch grid, no angulation at all is permitted.

Other causes of grid cutoff are a too long or a too short FFD (Figure 13-13), having the central ray off-center to the center line of the grid, and using a focused grid upside down (Figure 13-14).

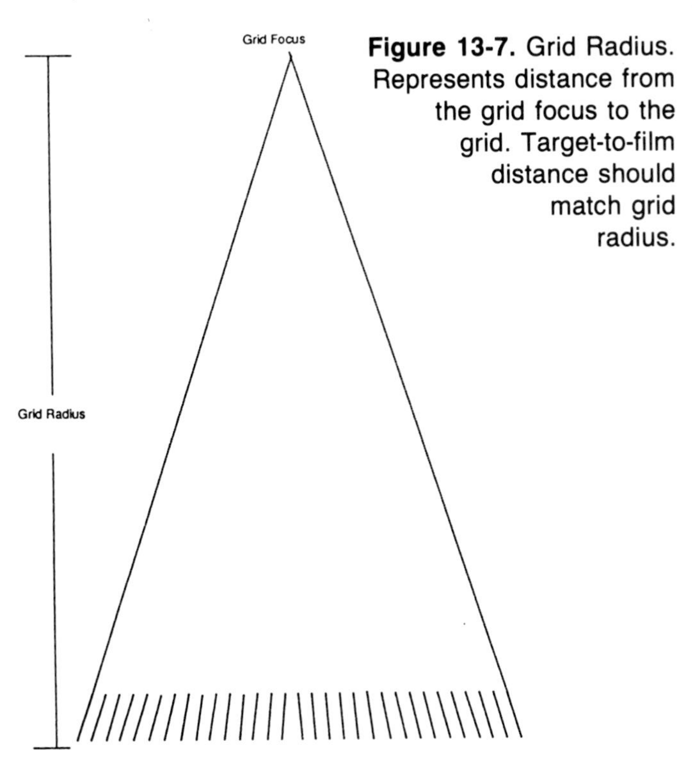

Figure 13-7. Grid Radius. Represents distance from the grid focus to the grid. Target-to-film distance should match grid radius.

Figure 13-8.

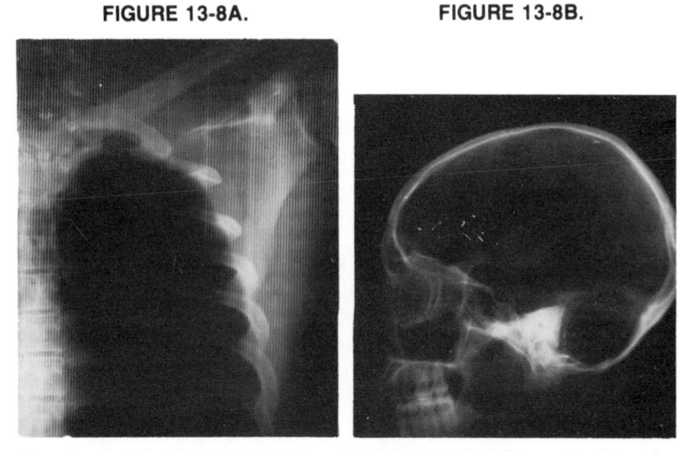

On the left in the scapula radiograph (13-8A), note the obvious vertical grid lines. In recent years technology improvements have given grid manufacturers the capability to produce grids with extremely narrow strips that are more closely spaced. Compare the much finer horizontal grid lines on the lateral skull radiograph (13-8B).

Figure 13-9. Photographs of Potter-Bucky Diaphragm. In the top photograph is a top view of a Potter Bucky diaphragm assembly with the x-ray tabletop removed. Note that the centerline of the grid lines up with the centerline of the table. In the lower photograph the grid has been removed and the film tray has been pushed to the closed position.

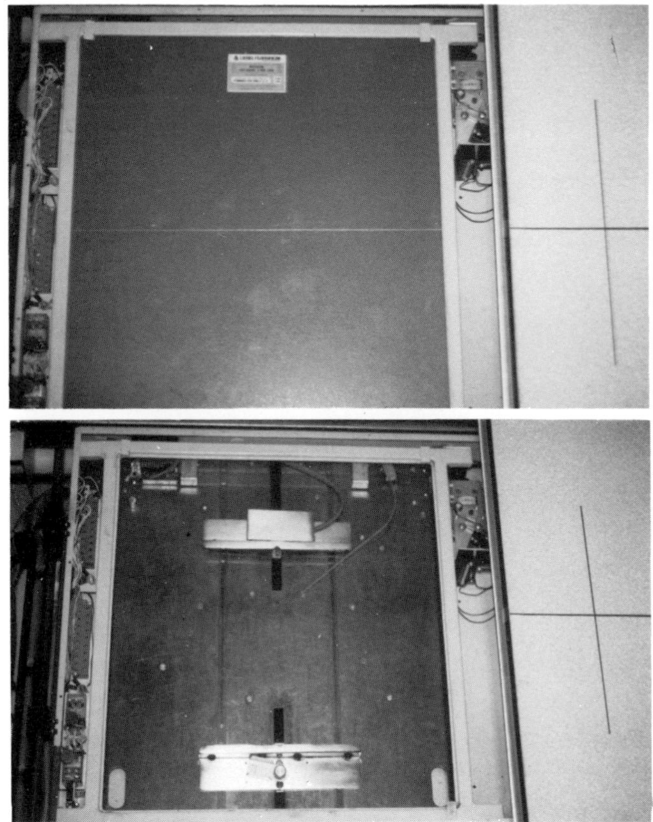

Figure 13-10. Potter-Bucky Diaphragm (end cross-sectional view). Movement of the grid from side to side is designed to eliminate the appearance of grid lines.

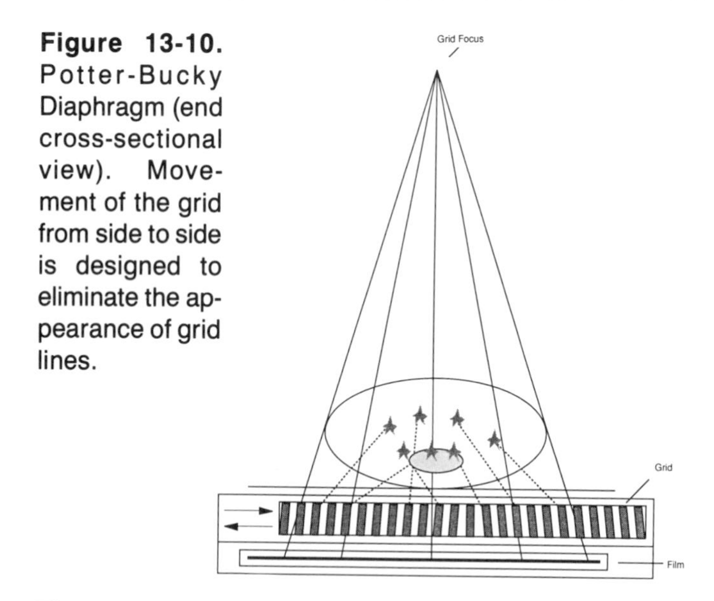

Figure 13-11. Potter-Bucky Diaphragm. The diaphragm is below the surface of an x-ray table. Since the lead strips in this device run longitudinally, angulation of the x-ray tube is possible, so long as this angulation is in the same direction (toward either end of the table.)

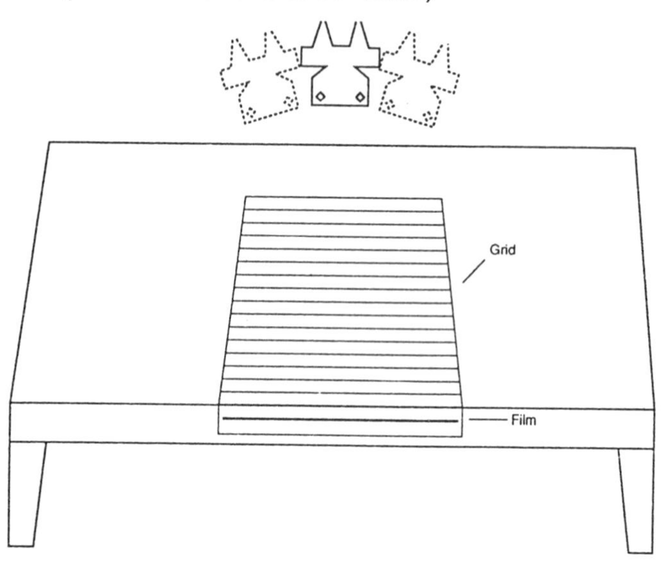

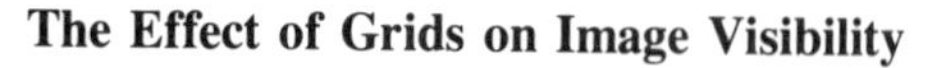

The Effect of Grids on Image Visibility

Radiographic Density

As has been stated, grids absorb both primary and scatter radiation. The higher the grid ratio, the greater the amount of radiation absorbed. As a result, radiographic density decreases as grid ratio increases. The relationship between grid ratio and density is therefore inverse (Figure 13-15).

Radiographic Contrast

Because scatter radiation appears on film as an extra unwanted grayness, anything that removes it will leave the film with more difference between the black-and-white tones, or in more technical terms, higher contrast. A higher contrast radiograph can also be referred to as having a shorter scale of contrast (fewer intermediate shades of gray). Use of a grid, therefore, will increase the radiographic contrast, and shorten the scale of contrast. The higher the ratio, the greater the radiographic contrast, and the shorter the scale of contrast. Changing from a higher grid ratio to a lower grid ratio will result in lower contrast on the radiograph, or an increased scale of contrast. The relationship is, therefore, direct between grid ratio and radiographic contrast (Figure 13-16). The relationship between grid ratio and scale of contrast is inverse.

Figure 13-12. Grid Cut-off Caused by Angling Across the Grid.

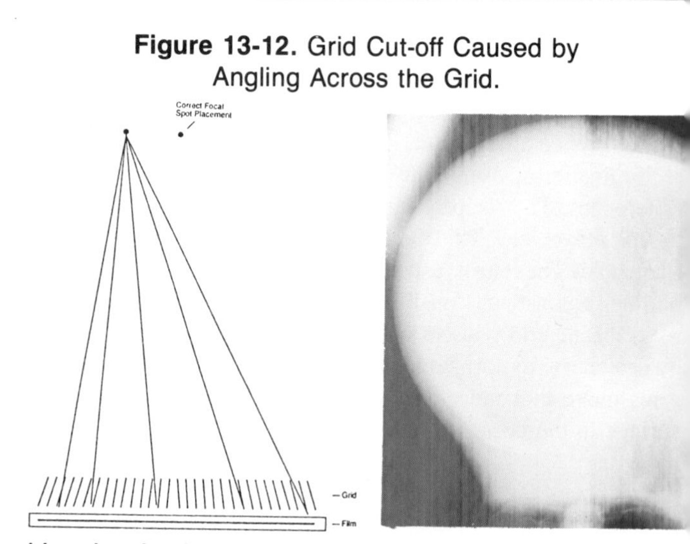

Line drawing illustrates the problem of angling across the grid. Lateral skull radiograph was taken with a 5 degree cross angle. Identical results can occur when x-ray tube is not positioned over the centerline of the grid even when using perpendicular cental rays.

Figure 13-13. FFD too short or too long. FFD must be selected to match the grid radius.

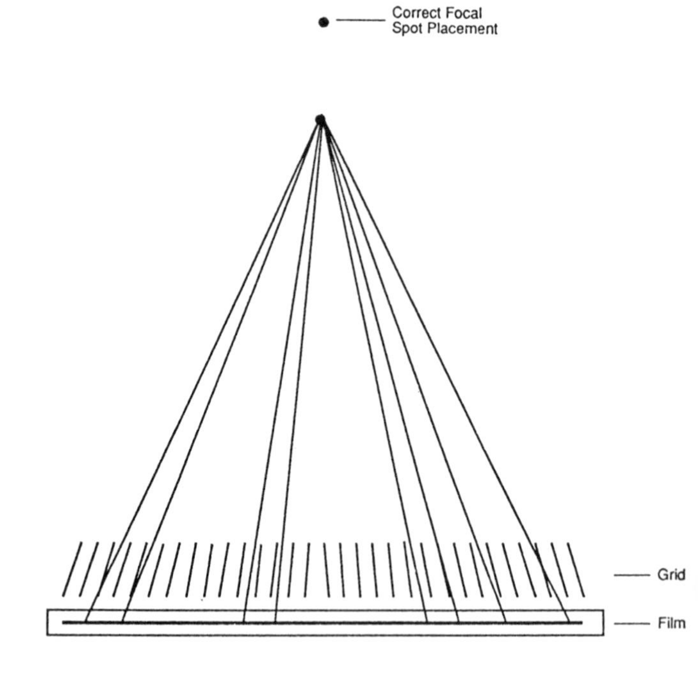

Figure 13-14. Wafer Grid Upside Down.

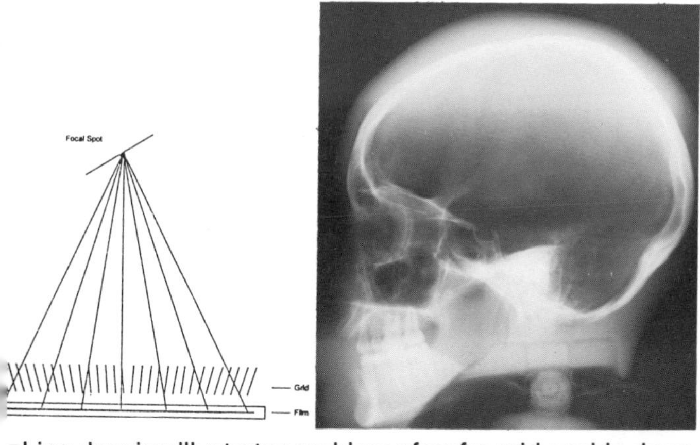

Line drawing illustrates problem of wafer grid upside down. Lateral skull radiograph was taken with wafer grid upside down. Similar radiograph will result from FFD mismatched to grid radius.

The Effects of Grids on Image Formation

Recorded Detail

While the use of a grid might improve the visibility of anatomical details, grids do *not* affect the recorded detail. One possible exception to this statement might occur when changing from tabletop to Bucky work which introduces some object-to-film distance into the situation. However minimal, the resulting loss of sharpness is from an OFD change and not the use of a grid.

Size and Shape Distortion

The preceding paragraph addresses the possibility that the use of a grid and resultant OFD would slightly magnify the image. However, it should be noted that any noticeable enlargement is due to an increase in object-to-film distance and not the grid. Therefore, the use of grids does not affect shape distortion.

SUMMARY

Grids are essential for removal of the large quantities of secondary and scatter radiation produced with some views. Grids consist of many small strips of lead separated by radioparent materials. The strips may be parallel to each other, as in an unfocused grid. The strips may be angled toward a point as in a focused grid, or the grid may consist of two sets of strips at right angles to each other as in a cross-hatch grid. Certain terms are important for understanding other differences among grids. Grid ratio is defined as the comparison of the height of the lead strips to the space between the strips. The number of lines per inch will be indicative of the visibility of the lead strips and the absorption of scatter radiation. The grid focus is the point above the focused grid where the lead strips would converge if extended into space above the grid. The grid radius is the distance between the grid focus and the grid. The target-to-film distance used should match the grid radius.

Grid lines will be seen when taking radiographs using wafer grids or grid cassettes. A wafer grid is a grid by itself. This kind of a grid is taped or otherwise secured ito the front of a regular filmholder. A grid cassette is a filmholder with a grid built into the front surface of the cassette. The use of a Potter-Bucky Diaphragm is required if grid lines are not to be seen. The Bucky, as it is more common called, is a mechanism that is built beneath the x-ray table surface. During the exposure, the mechanism causes the grid to move with a rapid, reciprocating, side-to-side motion that has the effect of blurring the appearance of the grid lines. Grid cutoff is the effect on film if the grid starts to unevenly absorb primary radiation. Grid cutoff has four causes: 1) angling across the grid, as opposed to angling with the grid. (No angulation at all can be used with a cross-hatch grid.) 2) using too long or too short a FFD. 3) having the CR off center to the grid center. 4) having a focused grid upside down.

Grids do have an effect on radiographic density and contrast.

Figure 13-15.

FIGURE 13-15A. **FIGURE 13-15B.** **FIGURE 13-15C.** **FIGURE 13-15D.**

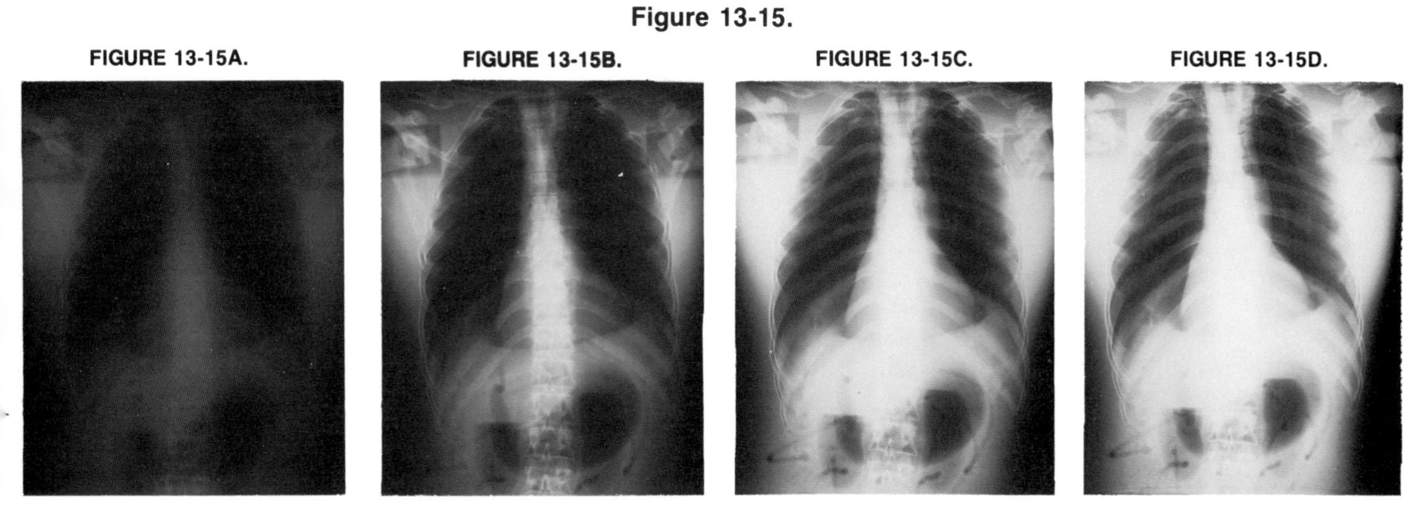

All four views are taken on the same patient (a radiographic phantom) with identical x-ray machine factors and film/screen combinations. The radiograph in Figure 13-15A is taken non-grid as indicated by the greater density and fog. The radiograph in Figure 13-15B is taken with a 5:1 grid; 13-15C with a 8:1 grid; and 13-15D with a 12:1 grid. Higher grid ratios result in less density all other factors being equal.

All three views are taken on the same patient (a radiographic phantom) with identical film/screen combinations and x-ray machine factors adjusted to produce equivalent radiographic densities. The radiographic in Figure 13-16A is taken non-grid as indicated by a large amount of fog. The radiograph in Figure 13-16B is taken with a 5:1 grid and has noticeably higher contrast than 13-16A; 13-16C with a 12:1 grid has the greatest amount of radiographic contrast. An increase in grid ratio therefore results in higher contrast.

Figure 13-16. Radiographic Contrast as Affected by Grids.

FIGURE 13-16A. FIGURE 13-16B. FIGURE 13-16C.

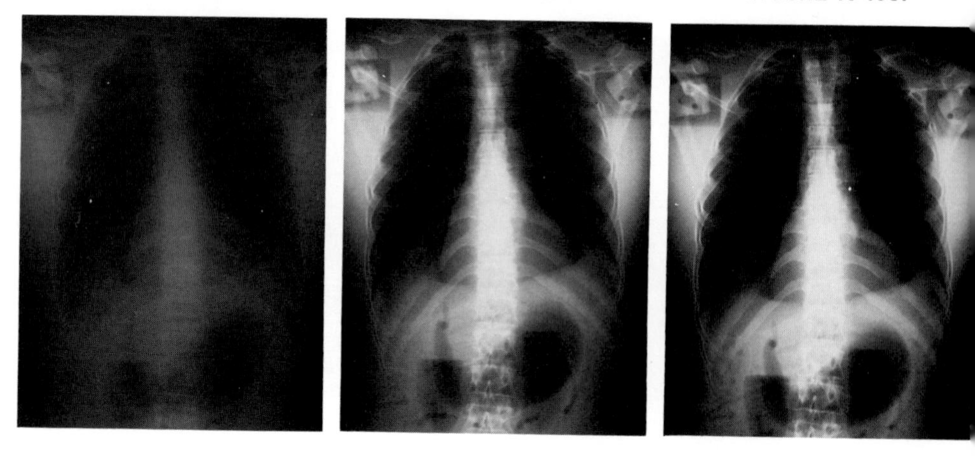

Grids of a higher ratio will reduce density and increase contrast (or shorten the scale of contrast). Lower ratio grids will produce radiographs of greater density and less contrast, or lengthen the scale of contrast. Grids do not have an effect on recorded detail, size or shape distortion. A summary of these radiographic effects is shown in Table 13-2.

TABLE 13-2. **Summary of the Effects of Grid Ratio on Radiographic Quality.**

IMAGE VISIBILITY	
Radiographic Density	–
Radiographic Contrast	+
IMAGE FORM	
Recorded Detail	0
Size Distortion	0
Shape Distortion	0

+ = Direct Relationship to Image Characteristic
– = Inverse Relationship to Image Characteristic
0 = No Effect on Image Characteristic

STUDY QUESTIONS

1. What is the role of a grid in radiography?
2. Compare parallel, focused, and cross-hatch grids.
3. Define grid ratio, grid focus, and grid radius.
4. Compare wafer grids and grid cassettes.
5. How can the appearance of grid lines be eliminated from a radiograph?
6. List four possible causes of grid cutoff.
7. If the grid ratio is changed from 5:1 to a 12:1, what, if anything, will happen to radiographic density; to radiographic contrast; to the scale of contrast?
8. If the grid ratio is changed from a 10:1 to a 6:1, what, if anything, will happen to recorded detail?
9. If the grid ratio is changed from a 5:1 to a 12:1, what, if anything, will happen to size distortion; to shape distortion?
10. Given the three radiographic situations shown below in A, B, and C, identify the one producing: a) the most density; b) the least density; c) the highest contrast; d) the lowest contrast; e) the longest scale of contrast; f) the shortest scale of contrast (all factors not listed assumed equal).

A	B	C
10:1 grid	5:1 grid	16:1 grid

BIBLIOGRAPHY

1. Bushong, Stewart C. *Radiologic Science for Technologists*, 2nd ed. St. Louis: C. V. Mosby Company, 1980.
2. Cahoon, John B. *Formulating X-Ray Techniques*, 8th ed. St. Louis: C. V. Mosby Company, 1980.
3. *Characteristics and Applications of X-Ray Grids*. Cincinnati: Liebel-Flarsheim, 1983.

Chapter 14

The Effects of Intensifying Screens on Radiographic Quality

Chapter Outline

1. Learning Objectives
2. Introduction
3. The Role of Intensifying Screens in Radiography
4. The Effects of Intensifying Screens on Image Visibility
   A. Radiographic Density
   B. Radiographic Contrast
5. The Effects of Intensifying Screens on Image Formation
   A. Recorded Detail
   B. Size Distortion
   C. Shape Distortion
6. Summary
7. Study Questions
8. Bibliography

Learning Objectives

Upon completion of this chapter, the student should be able to:

1. Draw a cross section of an intensifying screen, and label its different layers.
2. Explain the function of each layer of an intensifying screen.
3. List the characteristics that phosphor crystals need to possess for use on an intensifying screen.
4. Identify the advantages and disadvantages of intensifying screens.
5. Explain how intensifying screens affect the following:
   A. radiographic density
   B. radiographic contrast
   C. recorded detail
   D. distortion

Introduction

The advent of intensifying screens came during the infancy of radiography. Scientists were already experimenting with different materials that fluoresced, and with Roentgen's discovery of x-rays, research increased into the study of and experimentation with fluorescent compounds. This research gave birth to the efficient use of fluoroscopic intensifying screens. Intensifying screens are a valuable tool in the radiography profession, and the radiographer must know the screen's characteristics, advantages and disadvantages, and construction.

The Role of Intensifying Screens in Radiography

All intensifying screens are basically constructed in a similar way. The major difference is the type of phosphor crystals that the screen possesses. Figure 14-1 illustrates a cross section of an intensifying screen. The support layer/base is made of paper, plastic, or cardboard and gives the screen the required amount of stiffness. This layer will be coated with the fluorescent crystals. To help keep this screen from warping, a curl-control backing is sometimes applied to the base layer.

The next layer is the undercoat or reflective layer. When photons hit the fluorescent crystals, these crystals will fluoresce and light will be emitted both forward and backward. Normally, the light being emitted backward is lost; however, the placing of a reflective layer behind the crystals will redirect the light toward the x-ray film. (Note: Not all screens have this layer.)

On top of the reflective layer is the phosphor layer with tiny crystals. The type, size, and thickness of the crystals can vary. Because of these variables, intensifying screens have different speeds, which will be discussed later in this chapter.

The final layer is a protective coating. This coating protects the crystals from abuse such as handling of screens with dirty

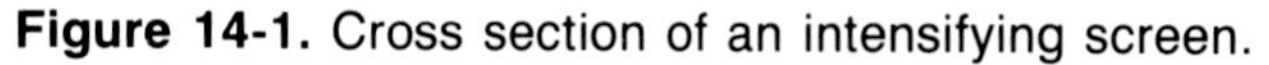

Figure 14-1. Cross section of an intensifying screen.

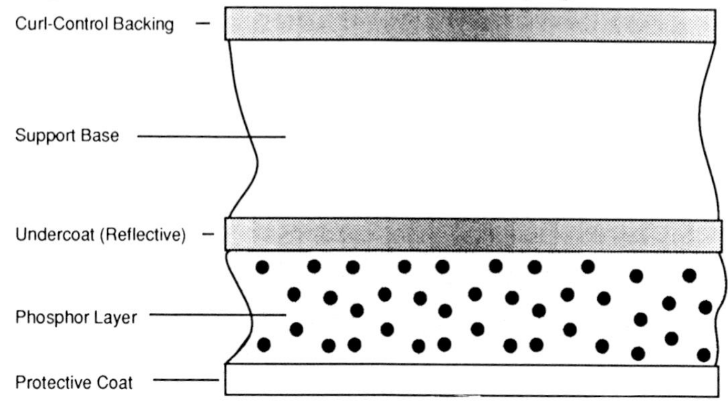

hands. This protective coating on the intensifying screens permit the cleaning of the screens with a specific cleaner or soap and water, and also protects the phosphor layer from scratches and abrasions.

The heart of the intensifying screen is the phosphor crystals layer. The purpose of the phosphor crystal is to absorb the photons (x-rays), convert the x-ray energy to light and emit light photons. The amount of light emitted is proportionate to the amount of x-ray photons absorbed by the crystals. In other words, photon energy is converted to a light energy which affects the film.

The characteristics of the phosphor crystals will determine the efficiency of the screen. All crystals that have ability to fluoresce are not suitable for intensifying screens. Crystals must possess certain properties to be useful in the field of radiology. To be effective these crystals must respond over a wide range of kV, must not deteriorate with age, must have minimal after-glow (screen lag), and must be in the visible light emission spectrum that is most sensitive to the film being used.

The most common phosphor crystals that are used today in the manufacturing of intensifying screens are calcium tungstate, phosphor-containing barium, and phosphors from the lathanide or rare-earth series. Each one of these phosphors will emit a different wavelength or colored light and corresponding films are manufactured to be selective in their response to various wavelengths (color). Calcium tungstate, for example emits a blue violet color whereas lanthanide phosphor crystals emit a brighter bluish green color. Films sensitive to blue-violet light would then be coupled with the calcium tungstate screen, and film sensitive to the bluish-green spectrum would be paired with the latter phosphor.

Intensifying screens are rated according to speed. The factors affecting speed are the inherent film speed, the absorption efficiency of the screen, the screen spectral emission match with film-spectral sensitivity, and screen efficiency in converting absorbed x-ray to light.

Inherent film speed is a built-in characteristic of diagnostic x-ray film. All radiographic film does not have the same speed. Speed refers to the sensitivity of the film to exposure by x-rays or light. The faster the speed of the film the less exposure will be required to produce an acceptable radiograph. One can determine the film speed by the use of a graph called a characteristic, sensitometric, or H & D curve. This visual line graph is produced by plotting the density/exposure relationship.

Absorption efficiency of the screen relates to how efficient the phosphor absorbs the photons. This is influenced by two factors. One of these factors is the type of phosphor being used. Different phosphor crystals will absorb more x-ray photons than others. The second factor is the phosphor layer thickness. Increasing the thickness of the phosphor layer will increase the amount of phosphor crystals that are available to absorb the x-ray photons. Consequently, the thicker the phosphor layers the more absorption of the x-ray photons.

As previously stated, the screen spectral emission must match the film spectral sensitivity. If the screen emits a blue-violet or bluish-green color, the film must be compatible to that particular color.

TABLE 14-1. **Table of Different Factors Affecting Speed of Screens.**

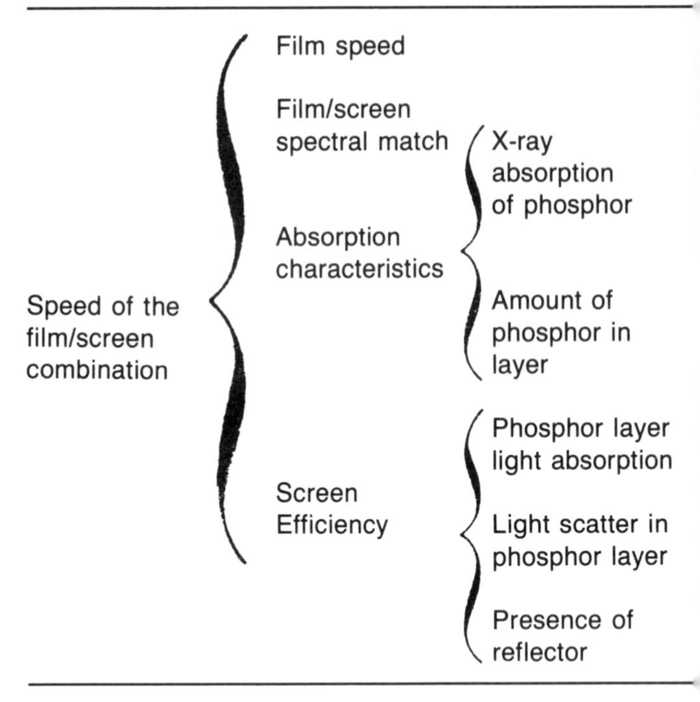

Screen efficiency indicates the efficiency of converting the absorbed x-rays to light. This is determined by the ratio of the energy of the light emitted by the screen to the x-ray energy absorbed. Screen efficiency is influenced by whether the screen has a reflective layer and the amount of light diffusion within the phosphor layer.

Table 14-1 presents the different factors which determine the screen speed. Contrary to some beliefs, intensifying screen speed is determined by more than the thickness of the phosphor crystals, the final determination of intensifying screen speed requires a compatible relationship between several different factors.

In general, intensifying screen speeds can be classified according to fine detail, par, high, and rare earth. These terms are very broad classifications, but are usually accepted in radiography. This screen classification represents the efficiency of converting x-radiation into usable light photons.

One can readily see the advantages of using screen filmholders over nonscreen filmholders. Nonscreen filmholders require the x-ray photons to do all the work in exposing the x-ray film. Since the phosphor crystals on intensifying screens absorb more photons than the radiographic film and give off considerably more exposure energy in the form of light, less photons are required to achieve the same results. Figure 14-2 demonstrates an anterior-posterior view of the knee taken on a manikin. Radiograph A used screen film and screen filmholder. The factors used were 60 kVp, 1.5 mAs, 40″ FFD, high-speed screens. Radiograph B, the same view, used screen film and nonscreen filmholder, and repeated the same technical factors. The quantity of photons be-

Figure 14-2.

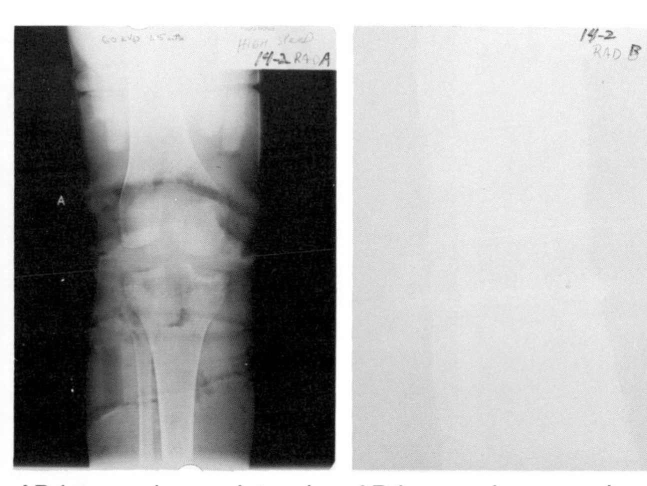

AP knee using an intensifying screen.

AP knee using a cardboard filmholder.

ing produced and striking the film were identical; hence, the only difference was the use of the intensifying screens. This should illustrate that an acceptable image can be produced with less x-radiation with the use of screen filmholders.

Another advantage in using intensifying screens is that lower exposure factors may be used. The patient therefore will receive less radiation. Figure 14-3 demonstrates an anterior-posterior view of a manikin's knee. Radiograph A was taken using a high plus speed screen and screen film. The technical factors were 60 kVp, 1.5 mAs, 40″ FFD. In Radiograph B, the same manikin's knee was radiographed using a nonscreen filmholder and screen film. The technical factors were adjusted to obtain the same amount of density on the radiograph. The factors for Radiograph B were 60 kVp, 300 mAs, 40″ FFD. The difference between the two techniques is that approximately two hundred times more mAs is required when using a nonscreen filmholder. With a decrease of exposure, shorter exposure times can be used. A short exposure time implies that the part being radiographed has less opportunity to move during the exposure.

Using lower technical factors will also result in the production of less tube heat, thus increasing the tube life. Additionally, a decrease in tube load allows a small, rather than a large, focal spot to be a safe choice in more examinations. Use of a small focal spot will improve the geometric sharpness.

Along with the advantages come the disadvantages. The use of an intensifying screen rather than a nonscreen filmholder will cause the image to be less sharp. The structural lines will not be as clearly defined as campared to nonscreen filmholders because of the thickness of the phosphor crystals. The thicker phosphor crystal layer causes a larger penumbra (unsharpness) to be produced. This will be explained later in this chapter.

Another disadvantage of intensifying screens is the possibility of producing quantum mottle. Quantum mottle is the variation in density of a uniformity exposed radiograph that results from random spatial distribution of the x-ray quanta absorbed in the screen. X-rays are bundles of energy and are referred to as photons or quanta. Using a nonscreen filmholder, photons will strike the x-ray film in uniform fashion. The photons will be absorbed by the crystals on the film, causing uniform blackness on the film when processed. The use of intensifying screens requires fewer photons to accomplish the same results. When using high-speed screens with fast-speed film, the finished radiograph may have a spotty (mottle) look. There will be areas on the radiograph that will have more blackness than other areas.

An analogy to help illustrate quantum mottle and its effects is: If one is sowing grass seed in a given areas with a certain amount of seed, and the seed was sufficient to cover the area so that each seed could be placed side by side, given the ideal circumstance, the grass would grow and cover the entire area evenly. No one part of the area would have more grass than another. This relates to a direct exposure method where the x-ray photons expose the x-ray film. Given the same area to sow grass seed, but this time there is only one-fourth the amount of seed. The seed is sowed so it will cover the area but some parts of the area will have more seeds than others, under the ideal circumstances, the grass will grow covering the required area. In some parts of the area there will be more grass than other parts. This is the type of situation that occurs when using intensifying screens of very high speed.

Figure 14-4 demonstrates the results of quantum mottle on a radiograph. Radiograph A was obtained from the direct exposure method using a nonscreen filmholder and screen film. After the filmholder was exposed to x-rays, the film inside was processed. Notice that there is a uniform radiographic density present throughout the film. Radiograph B was obtained from a high-speed screen and screen film. After the cassette was exposed and the film processed, the radiographic density was altered. Areas of the radiograph have a spotty appearance and some areas ap-

Figure 14-3.

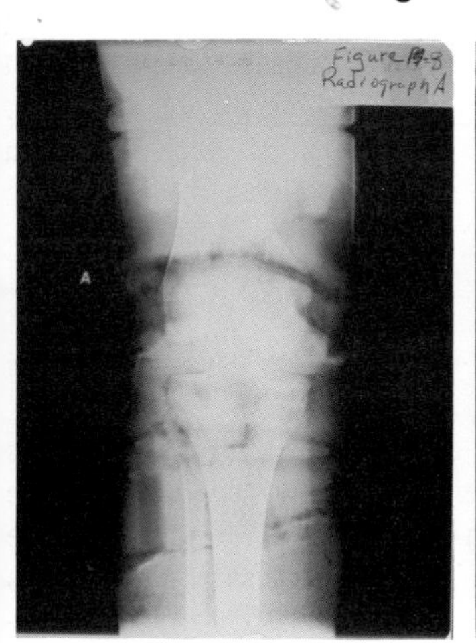

AP knee using high-speed screen 60 kVp, 1.5 mAs.

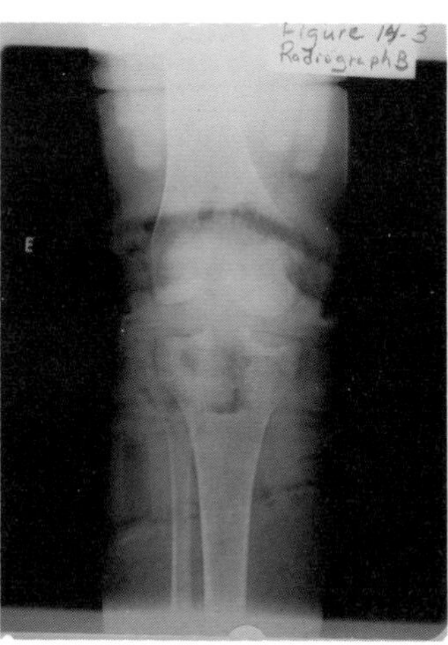

AP knee using a cardboard filmholder 60 kVp, 300 mAs.

Figure 14-4.

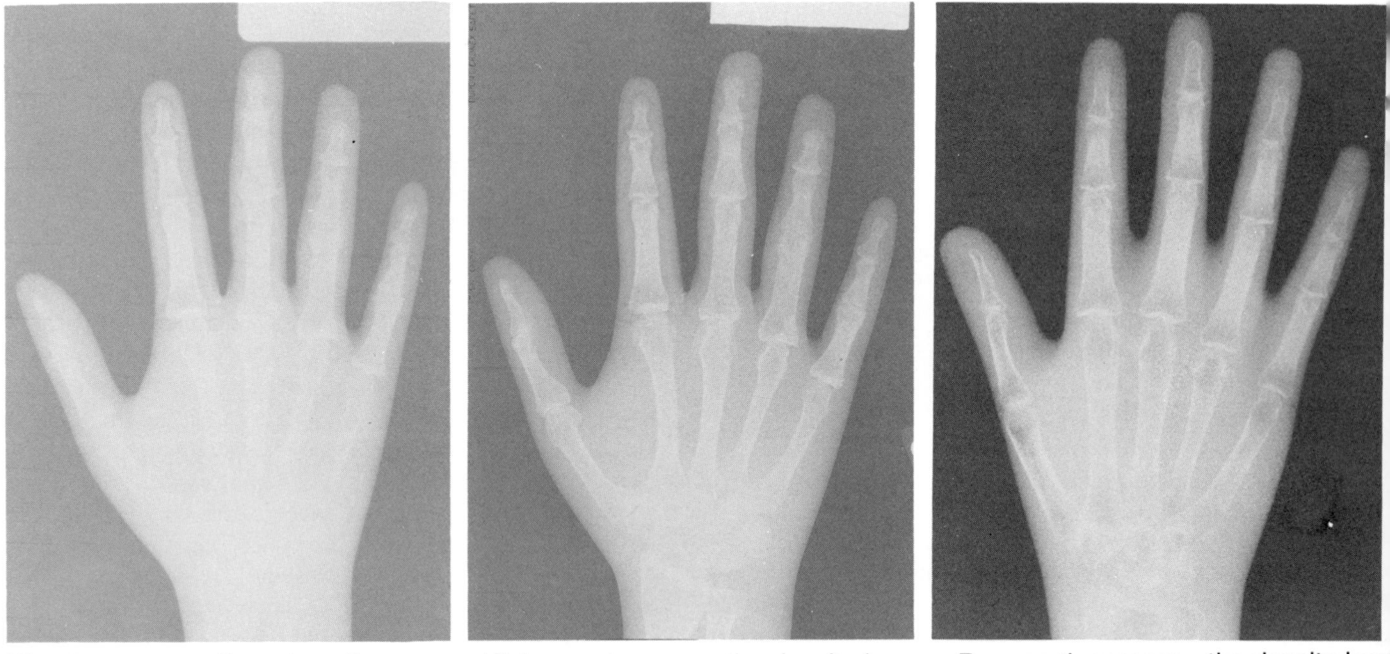

Direct exposure—there is uniform density throughout radiograph.

High-speed screen—the density has been altered and has a spotty appearance.

Rare-earth screens—the density has more of a spotty appearance.

pear more dense than others. Radiograph C was obtained using a rare-earth screen and screen film. After exposure and processing a spotty appearance is more noticeable on this radiograph. This spotting of density is called mottle.

Normally, quantum mottle will not interfere with the diagnostic interpretation of the radiograph, but it may be distracting in some cases. Mottle becomes a problem when dealing with films that are going to be magnified or enhanced. With magnification and enhancement the mottle affect increases. Quantum mottle increases with the speed of the intensifying screens. In Figure 14-4 notice that as the mAs is decreased because the speed of the screen increased, the quantum mottle also increased.

The Effects of Intensifying Screens on Image Visibility

Radiographic Density

A change in intensifying screen speeds will change the amount of density recorded on a radiograph. For example, if factors are the same, a change from detail to high-speed screens will increase the amount of density on the film. Figure 14-5 demonstrates three different size crystals and no crystal (nonscreen). Comparing the crystals one can see the size will affect the thickness of the crystals. The size of crystals include the height, width, and depth. Consequently, the size and the thickness can be considered the same. Size/thickness of the phosphor crystals help to determine the speed of the intensifying screen. (Remember the purpose of the phosphor crystal is to absorb the photons then fluoresce the compatible color for the x-ray film.) The larger the crystal the more absorption of the x-ray photon. An increase in the amount of photon absorption will cause the phosphor crystal to emit more light, creating more density on the film. As the crystals decrease in size/thickness the photon absorption decreases. Also, the light emitted by the crystals and the density on the film decreases.

Figure 14-6 demonstrates a posterior-anterior view of a manikin's hand. Four radiographs were taken using different screen speeds and one radiograph was taken without using

Figure 14-5. The effect of crystal size on density and penumbra effect.

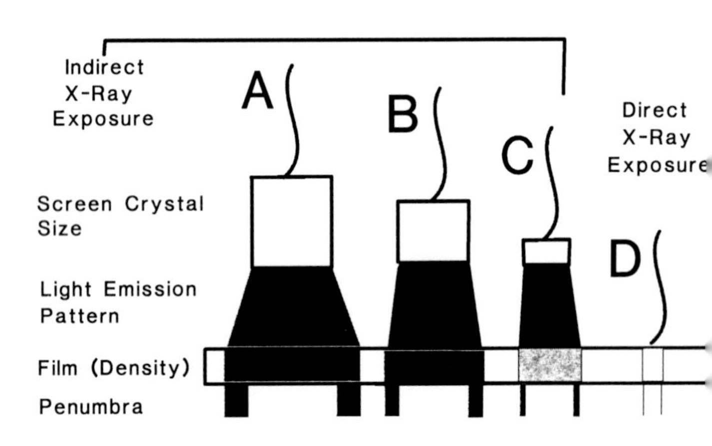

Different size/thickness crystals of intensifying screens and their affect on the speed. Crystal A emits more light (high speed); Crystal B emits medium light (par speed); Crystal C emits less light (slow speed); and D represents no crystal (direct exposure).

Figure 14-6.

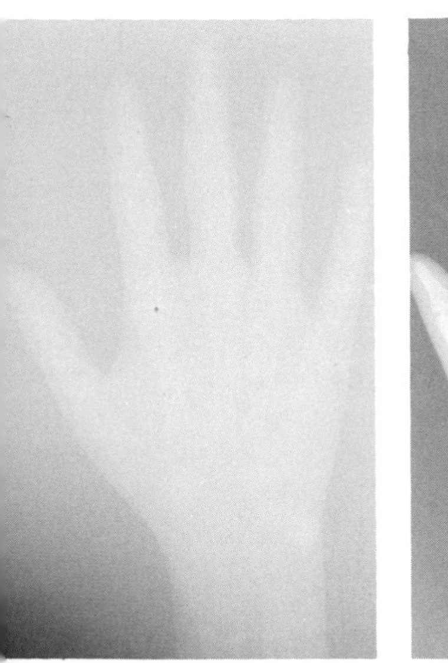

Cardboard filmholder. Film has no radiographic density.

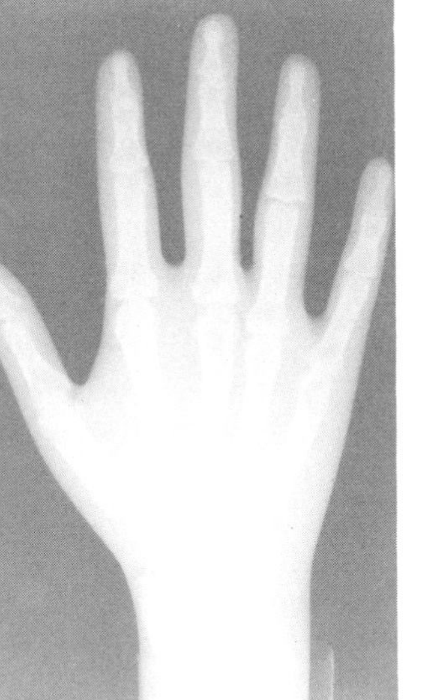

Par-speed screen. The radiographic density has increased.

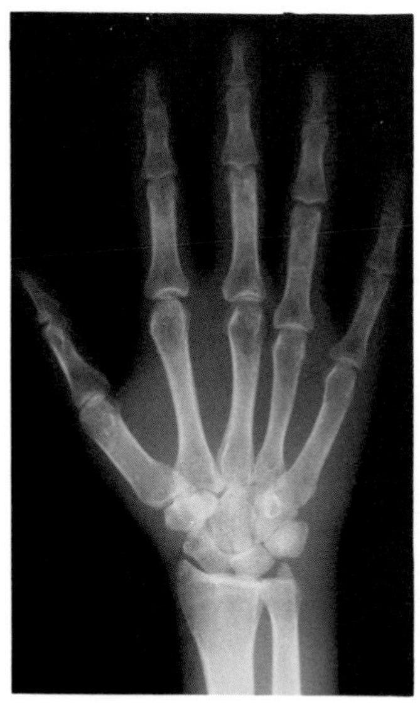

High-speed screen. Radiographic density sufficient.

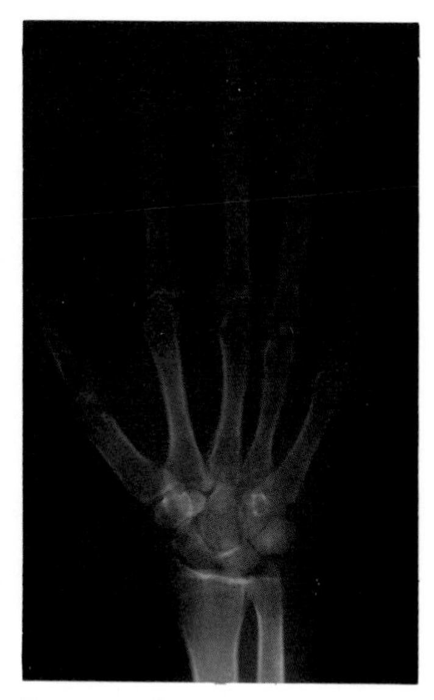

Rare-earth screen. Radiographic density is excessive.

screens. The technical factors used for all radiographs were 50 kVp, 2.5 mAs, 40″ FFD. Notice in Radiograph A that the film lacks sufficient density; consequently, the hand cannot be seen well. In Radiograph B, the density increased; however, it is not sufficient for a good-quality radiograph. Radiograph C was obtained using a high-speed screen. The amount of density in the film is sufficient for a good-quality diagnostic radiograph. Radiograph D used a rare-earth screen which has the ability to absorb three to four times the number of x-ray photons than the high-speed screen. Notice that the density here is excessive.

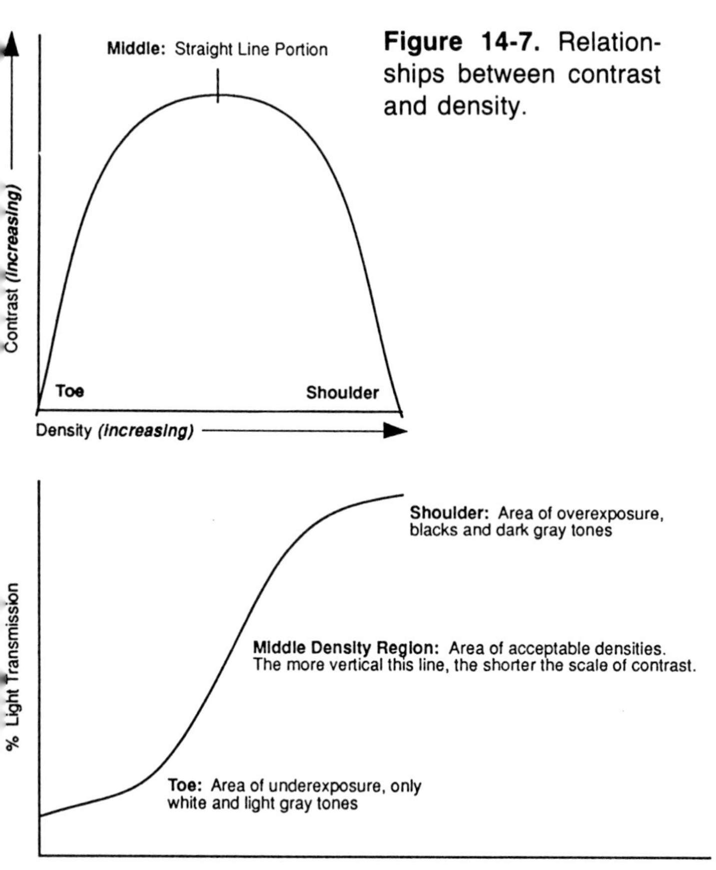

Figure 14-7. Relationships between contrast and density.

Radiographers need to know the relationship between density and the intensifying screens. If a change of intensifying screens is required, one must be able to adjust accordingly.

Radiographic Contrast

Film contrast changes with photographic density. If a radiograph is underexposed, structures of interest will have insufficient contrast because their density is low. This density is represented at the toe of the H & D curve. When a radiograph is overexposed, structures of interest will have insufficient contrast because their density is so high that density differences are minimal. Placing this density on the curve would represent the shoulder portion. The greatest visibility would be if the radiograph is exposed so that the structures of interest appear in the middle density region where film contrast is the highest. The two figures in Figure 14-7 explain this concept.

Different speeds of intensifying screens produce different contrast scales. Comparing, for example, the contrast of a radiograph using a high-speed screen to a radiograph using a par screen. Figure 14-8 is a comparison of these two screen filmholders. Both radiographs demonstrate an anterior-posterior view of a manikin's knee. Radiogaph A used the technical factors of 60 kVp, 1.5 mAs, 40″ FFD, high-speed screen and screen film. Radiograph B used the technical factors of 60 kVp, 6 mAs, 40″ FFD, par speed screen, and screen film. Notice the only technical factor changed was mAs, which was altered appropriately to produce

Figure 14-8.

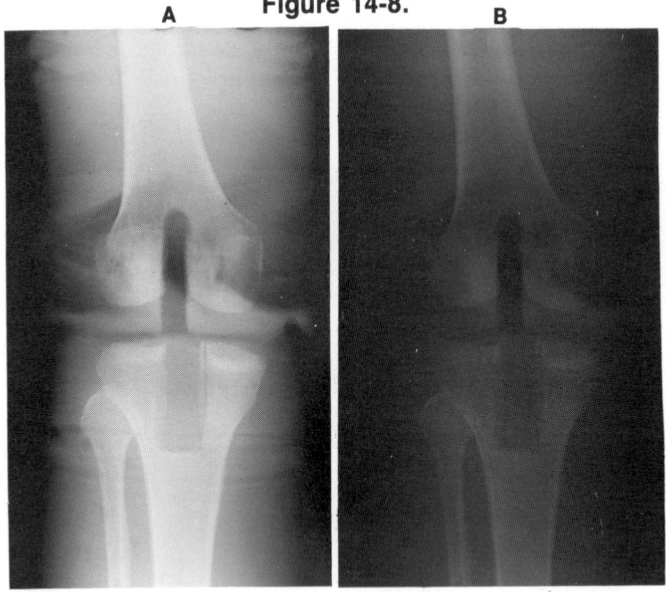

A: AP knee using high-speed screen 60 kVp, 1.5 mAs. Film has higher contrast than radiograph B.

B: AP knee using par-screen 60 kVp, 6 mAs. Film has lower contrast than radiograph A.

an image of similar density. The contrast on Radiograph A is higher compared to Radiograph B. Generally, as the speed on the screen increases, and all other factors except mAs remain the same, the contrast of the radiograph increases. This is a direct relationship between the intensifying screen speed and contrast.

The Effects of Intensifying Screens on Image Formation

Recorded Detail

The thickness of the phosphor crystal not only affects the speed of the intensifying screen, but also the recorded detail. The thicker the phosphor crystal layer, the greater the distance from the phosphor layer to the film thus producing less sharpness on the radiograph. Remember, the larger the penumbra, the less recorded detail the radiograph will possess. Crystal A has the largest penumbra and crystal C has the smallest penumbra. Consequently, the faster the speed of the intensifying screen the poorer the recorded detail.

Figure 14-9.

A: Hand and objects radiographed using a high speed screen. Images lack sharpness compared to radiograph B.

B: Hand and objects radiographed using a cardboard film holder. Images are sharper than in radiograph A.

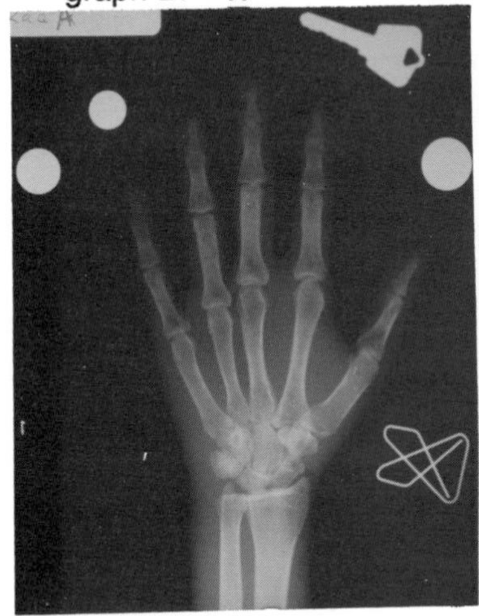

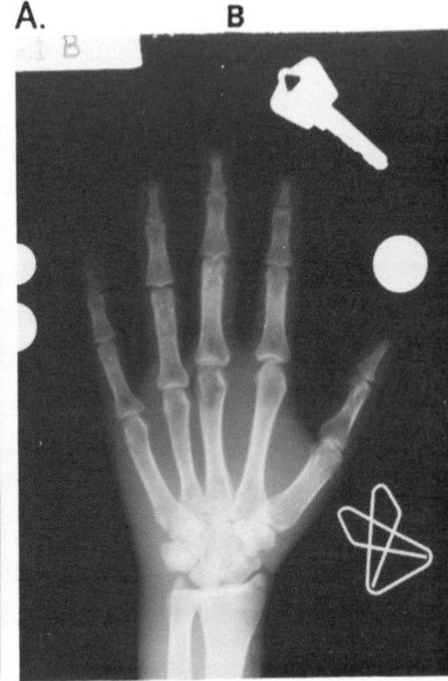

Figure 14-9 demonstrates two radiographs, one using a highspeed screen and one using a non-screen film holder. Both radiographs are of a posterior-anterior view of a manikin's hand. The technical factors have been adjusted for each so that densities are similar for both radiographs. Placed on the radiographs were coins and keys to help demonstrate the definition. Radiograph A used a high-speed screen and Radiograph B used a non-screen film holder. Compare the recorded detail between the two radiographs. Notice on Radiograph B that the structural line of the coins and the keys are well defined. On Radiograph A, the structural lines are defined, but not as well as on the preceding image. Notice that in Radiograph B, the teeth of the keys are easily defined and the ridges can be seen on the coins. In Radiograph A, the coins and key are imaged and have detail, but the teeth on the keys and the ridges on the coins cannot be seen. It can be concluded, then, that a radiograph taken on a non-screen film holder will have better recorded detail than one taken using intensifying screens.

If nonscreen filmholders and detail speed screens give the best recorded detail, why are they not used more frequently in the modern radiography department? Using different screen speeds has a trade-off. The slower screen has better recorded detail, but it requires more radiographic exposure to the patient. The high-speed screen demonstrates poorer recorded detail, but it requires less radiation exposure to the patient.

In the field of medical radiography today a major concern is the amount of radiation the patient receives. This is one reason why the newer and faster type of intensifying screens are chosen

TABLE 14-2. **Relative Speeds of Intensifying Screens.**

SCREEN SPEED	USE	RELATIVE SPEED
Detail or slow	Requires greater exposure but produces excellent detail, contrast, and sharpness.	1
Average or par	Requires moderate exposure and produces good sharpness and detail.	2
Fast or high	Requires very short exposure and produces slightly less sharpness and detail.	4

NOTE: The higher the relative speed number, the faster the speed. Higher relative speed numbers indicate faster screen speeds.

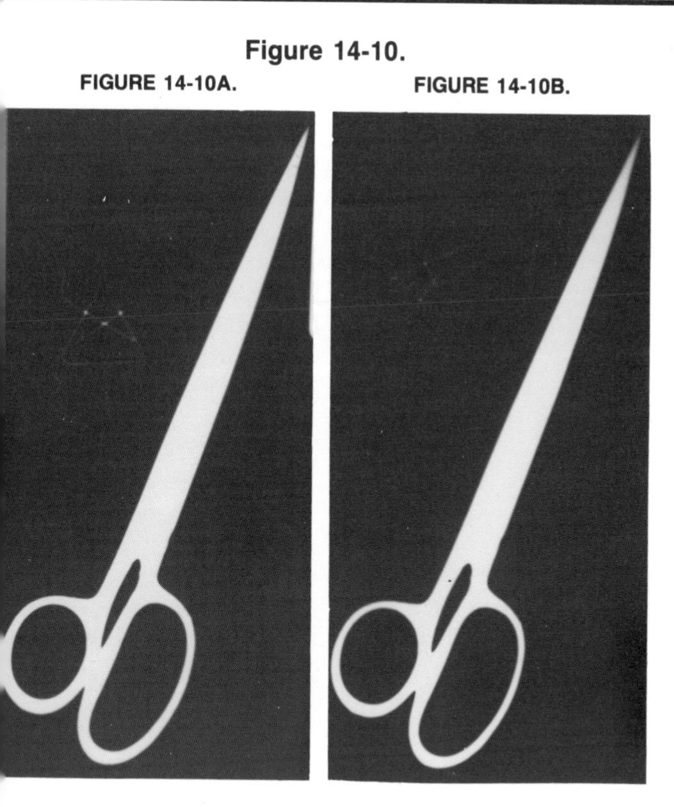

Figure 14-10.

FIGURE 14-10A. Jsing high-speed intensify-ng screens magnification s not observable.

FIGURE 14-10B. Using rare-earth screens magnification is not observable.

nstead of the nonscreen filmholders or slow-speed screens. The hart in Table 14-2 compares the speed of screens to the use of he screen and is a helpful guide to radiographers when they are aced with changing from one screen speed to aother.

ize Distortion

Size distortion/magnification is affected by geometric changes. t is true that the thickness of the phosphor does change the eometric arrangement between the crystal layer and the film, ut this change is not significant enough to produce observable nagnification. In Figure 14-10 a large paper clip and a pair of cissors were radiographed. Radiograph A used a high-speed creen and Radiograph B used a rare-earth screen. Comparing ne imaged items to the actual size of the items magnification is ot observable.

hape Distortion

To produce shape distortion on a radiograph the part, film or -ray tube must be nonaligned. Even though the screens are a art of the cassette, the angling of the screen by themselves will ot produce shape distortion. The screen filmholder must be an-led in relationship to the part being radiographed or to the x-y tube.

SUMMARY

Table 14-3 outlines the effects of intensifying screens on radiographic quality. If the speed of the screen is increased and all other factors remain the same, then the amount of density will increase on a radiograph. This is a direct relationship. The faster the screen, the less the exposure factors required to obtain the proper density on the radiograph and contrast will be high/short scale. The thicker the phosphor layer, the greater the distance between the screen and film. This will produce a larger penumbra causing greater geometric unsharpness. A faster screen speed calls for less photons, consequently fewer photons will cause quantum mottle on the radiograph. Intensifying screens with faster speeds allow a shorter exposure time which implies that motion on the radiograph could decrease.

Comparing all of the advantages and disadvantages of intensifying screens, the advantages of these screens outweigh the disadvantages. Without intensifying screens, radiographing some parts of the body would be almost impossible. Radiographers need to know, understand, and use intensifying screens to their fullest, as screens can significantly reduce the radiation dose to the patient.

TABLE 14-3. **Summary of Intensifying Screens on Radiographic Quality.**

IMAGE VISIBILITY	
Radiographic Density	+
Radiographic Contrast	+
IMAGE FORM	
Recorded Detail	–
Size Distortion	0
Shape Distortion	0

+ = Direct Relationship to Image Characteristic
– = Inverse Relationship to Image Characteristic
0 = No Effect on Image Characteristic

STUDY QUESTIONS

1. Label the different parts of an intensifying screen.
2. Name three characteristics that a phosphor crystal must possess in order to be used on an intensifying screen.
3. What is the function of the reflective layer on intensifying screens?
4. Name two advantages of using an intensifying screen.
5. Name a disadvantage of using an intensifying screen.
6. Explain how the use of intensifying screens will affect radiographic density.
7. Explain how the use of intensifying screens will affect radiographic contrast.
8. Explain how the use of intensifying screens will affect recorded detail.

BIBLIOGRAPHY

1. Bushong, Stewart C. *Radiologic Science for Technologists*. 2nd ed. St. Louis: C. V. Mosby Co., 1980.
2. Cahoon, John B. *Formulating X-Ray Techniques*. 6th ed. Durham, N.C.: Duke University Press, 1974.
3. Donohue, Daniel P. *An Analysis of Radiographic Quality*. Baltimore, Maryland University Press, 1981.
4. Fuchs, Arthur W. *Principle of Radiographic Exposure & Processing* Springfield, Ill.: Charles C. Thomas, 1973.
5. *Radiographic Noise*. Rochester: Eastman Kodak Company, 1982.
6. *Screen Imaging*. Rochester: Eastman Kodak Company, 1976.

Chapter 15

The Effects of Motion on Radiographic Quality

Chapter Outline

1. Learning Objectives
2. Introduction
3. The Role of Motion in Radiography
4. The Effects of Motion on Image Visibility
   A. Radiographic Density
   B. Radiographic Contrast
5. The Effects of Motion on Image Formation
   A. Recorded Detail
   B. Size
   C. Shape
6. Summary
7. Study Questions
8. Bibliography

Learning Objectives

Upon completion of this chapter, the student should be able to:

1. Define motion.
2. List two types of motion.
3. List three ways to eliminate motion.
4. Explain how motion can be useful in radiology.
5. Explain how motion affects the following:
   a. Contrast.
   b. Recorded detail.
   c. Magnification.
   d. Distortion.

Introduction

At one time or another, almost everyone has taken photographs that turned out blurred or fuzzy. One could not determine the object photographed because the structure lines of the image were poorly defined. The photograph, showing image details which were visible but unsharp, had poor recorded detail. Why did the photograph come out this way? Some may say the person holding the camera moved. Others may say the object being photographed moved. In any event, it is safe to assume that a blurry photographed was produced because of motion.

The same principle applies to medical radiography. If there is movement during an exposure, then the radiograph will be blurry or fuzzy. Generally, there will be a loss of recorded detail and the structural lines of image details may not be well seen. Remember, the radiologist makes a diagnosis of the patient from the finished radiograph. A radiograph that does not have adequate sharpness of detail may result in misinterpretation of the condition of the patient or may lead to a repeated examination.

The Role of Motion in Radiography

Motion may be defined as the act or process of moving something from one place to another. An example would be the act of moving the body or any of its parts. This example implies that the body or its parts has to move in order for motion to be observed. This, however, is not true. Motion can be caused by more than just the part moving.

Moving is the biggest single cause of loss of recorded detail on a radiograph (Selman, 1979). When motion is viewed on a radiograph, it appears as a blurring or fuzziness of the structures of the part being radiographed. A radiographer should routinely strive to produce a radiograph with the least amount of motion or no motion at all. In order to produce a radiograph with the best recorded detail, one must understand the nature of motion, the types of motion, and how to prevent motion.

Radiographers should be aware of two types of motion. The first can be classified as accidential and is the responsibility of the radiographer to eliminate this type of motion if possible. Accidental motion can be controlled by directions to the patient, short exposure times, part immobilization devices, proper breathing techniques, patient's cooperation, and ensuring that all x-ray equipment is locked in place.

Involuntary motion or physiological motion may be caused by respiration, heart action, muscle spasms, and tremors. The radiographer has minimal direct control over this type of motion, therefore, must assess the rate of respiration and peristalsis while establishing a short exposure time setting.

Motion in radiography has both positive and negative aspects. On the negative side, the radiograph will appear to be fuzzy or blurred if there is movement of the part, film, or tube during exposure. The structure lines will be unsharp with a loss of recorded detail, and depending on the rate of motion, there may even be a loss of image recognition. An analogy to this is a person waiting at a railroad crossing when a train passes by, but cannot readily make out the actual outline of the boxcars. Likewise, the letters on the boxcars are blurred and cannot be read. Motion on a radiograph would produce the same type of blurring effect. The structure lines of the part being radiographed would be unrecognizable. For this reason, radiographers should strive to produce a radiograph with the least amount of motion.

Looking at motion on the positive side, if one creates movement during an exposure to purposefully blur out the parts superimposing others, then radiographs could be produced for specific areas of interest. Radiographers use this concept in a wagging jaw maneuver when trying to demonstrate the second cervical vertebra. Figure 15-1 presents two radiographs of an anterior-posterior view of the cervical spine. Radiograph A shows that the mandible is obscuring the second cervical vertebra. Radiograph B demonstrates the use of the wagging jaw technique. Notice on Radiograph B that the structural details of the second cervical vertebra may now be visualized. Movement of the mandible during the exposure blurred the mandible radiographically and allowed the second cervical vertebra which did not move, to be visualized.

Radiographing the sternum is another example of when radiographers may use motion advantageously. Usually the patient is placed in a right anterior oblique position. The patient is instructed to begin shallow breathing during the exposure and to continue this breathing technique until requested to stop. This breathing during the exposure will cause movement of the ribs which in turn results in blurring of ribs radiographically. The sternum, which did not move, is clearly visualized on the radiograph.

The greatest use of motion in radiography is in the area of tomography. Tomography is a special technique in which various selected planes of the body can be clearly demonstrated on a radiograph while structures above or below are blurred in varying degrees. On a posterior-anterior view of the chest for example, all of the structures of the chest are shown in relation to each other; however, all these structures are superimposed upon each other. A radiologist may suspect the patient of an abnormality which is difficult to visualize. By using tomography, the radiologist is often able to determine the identity and location of the abnormality.

Figure 15-1.

AP cervical spine with mandible obscuring second cervical spine.

A

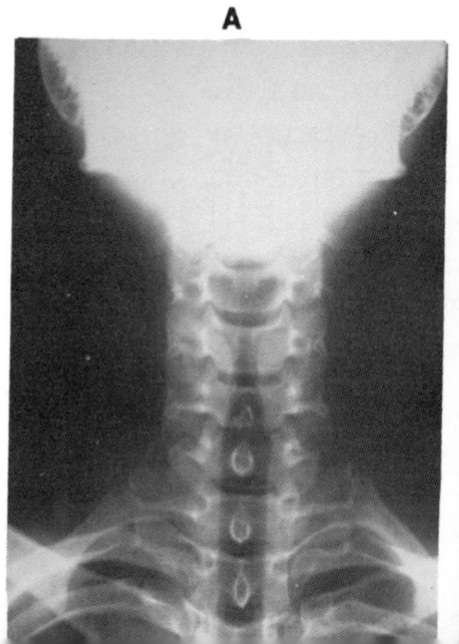

AP cervical spine with the wagging jaw technique.

B

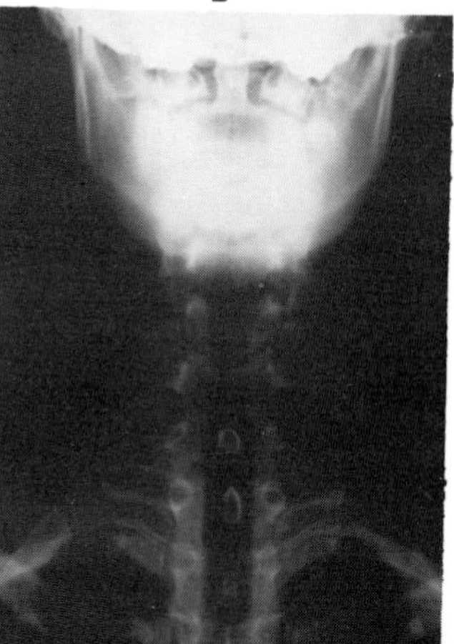

Tomography has many names like plainography, body-section, nephrography, and laminography. All of these work on the same principle. During the exposure the x-ray tube is moving in one direction while the film simultaneously moves in the opposite direction. The film and x-ray tube travel in the same plane that is regulated by a special fulcrum apparatus attached to the x-ray table and overhead tube. All anatomical structures being radiographed at the desired plane will appear sharply defined and well visualized. By changing the arc of travel of the x-ray tube, one can change the plane of interest within the part of the patient being radiographed.

The type of radiograph produced in tomography would be similar to the visual effect of a fast-moving train passing a crossing: the boxcars and the writing on the boxcars are blurred and difficult to visualize. As the boxcars pass by, however, one can look across to the other side of the tracks and see the automobiles waiting on the other side. These automobiles are not blurred but are very clear and well visualized. On tomographic radiographs, only the planes above and below the plane of interest will be ill-defined and obscured. The plane of interest, however, will be visualized with optimum definition of the structural lines.

Figure 15-2 shows four radiographs of a chest. Radiograph A is a routine view of a posterior-anterior chest. Although one can see the lungs, ribs, heart shadow, and the costophrenic angles, some of these structures are superimposed. If an abnormality is located in the lung area, it may be very difficult to identify and thus make an accurate diagnosis. Radiographs B, C, and D show tomograms of a chest using different planes of interest so the area that contains the abnormality can be better visualized at one plane as compared to another. The radiologist can locate the abnormality to determine its anterior or posterior position. Note that the blurring of the structures above and below the plane of interest in Radiograph B have changed in Radiograph C and D, and the structures that were once blurred have become defined.

The Effects of Motion on Image Visibility

Radiographic Density

Motion does not have an influence on the production of electrons or the speed of electrons in the x-ray tube. Therefore, motion will not change the quantity of x-rays in the primary x-ray beam. Generally speaking, motion does not have an affect on radiographic density. Figure 15-3 demonstrates this concept. Radiograph A shows an anterior-posterior view of both feet of a manikin. A technique of 50 kVp, 5 mAs, 40″ FFD was used. In Radiograph B the same technique was used, only the manikin's feet were moving during the exposure. The two radiographs illustrate the same amount of density. The only difference between Radiograph A and Radiograph B is that Radiograph B has a loss of recorded detail caused by movement of the part during the exposure.

Radiographic Contrast

Motion does not control the penetration of the primary x-ray

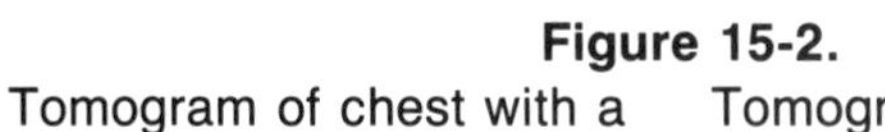

Figure 15-2.

A: Routine view of PA chest structures overlying each other.

B: Tomogram of chest with a plane of interest of 7 centimeters.

C: Tomogram of chest with a plane of interest of 10 centimeters.

D: Tomogram of chest with a plane of interest of 13 centimeters.

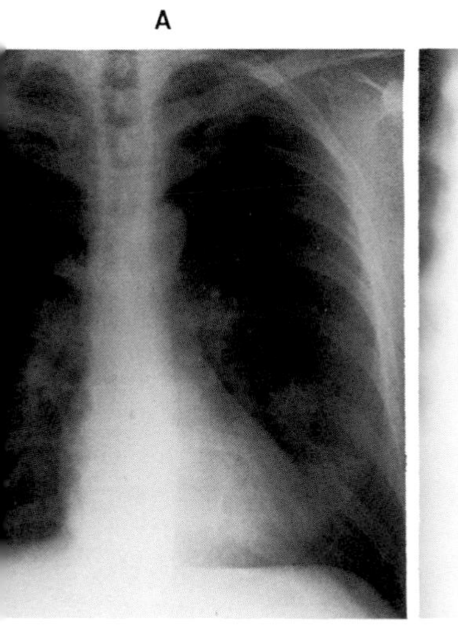

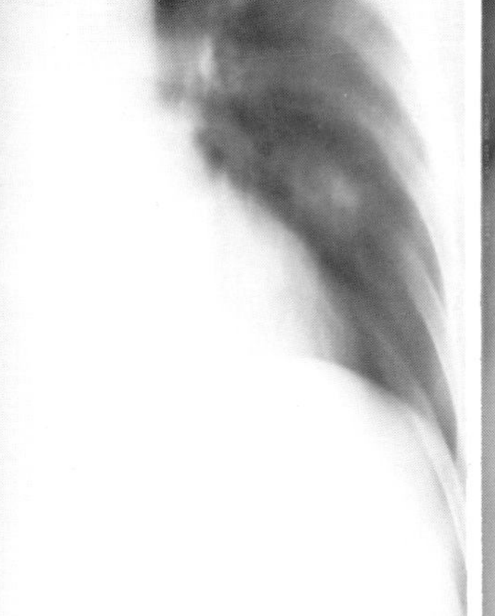

beam, but it should be remembered that the quality of the x-ray beam significantly affects radiographic contrast. Motion, therefore, has no effect on radiographic contrast. In Figure 15-4 one can compare the contrast between Radiograph A and Radiograph B. Both radiographs demonstrate an anterior-posterior view of an ankle. Radiograph A used factors of 60 kVp, 5 mAs, 40″ FFD; Radiograph B used identical factors, but during the exposure the ankle was moving. When comparing the two radiographs, one can see that contrast is unchanged. The only difference is that the structural lines defining the ankle on Radiograph B are not well-defined.

The Effects of Motion on Image Formation

Recorded Detail

When radiographers talk about recorded detail, they are referring to the sharpness of the structural lines of the image details on the radiograph. When viewing radiographs, radiographers should ask themselves if the structural lines are free from any blurring or fuzziness, and if everything was done to minimize motion caused by heart action, respiration, muscle spasms, and tremors. If there is a loss of recorded definition on the radiograph the radiographer should review the possible actions that can be taken to rectify the situation.

One of the best ways to reduce motion on radiographs is to gain the confidence of the patient. Patients coming into the radiography department are often frightened, confused, or depressed. The radiographer should cultivate the skills of effective communication. Explain to patients exactly what is going to happen to them, what will be expected of them, ask for their cooperation, and answer questions about the procedure prior to the actual examination. Make sure the patient understands the proper breathing technique by practicing the technique before making the exposure. Remember, ask the patient if he or she has

Figure 15-3.

A: AP view of both feet with no motion. The density the same as on radiograph B.

B: AP view of both feet with motion. The density the same as on radiograph A.

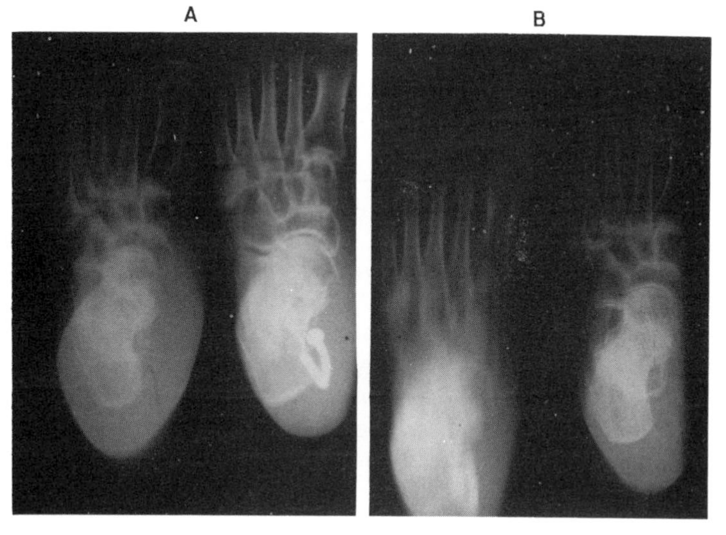

Figure 15-4.

A: AP view of the ankle using same technique as radiograph B. No change in contrast. No motion during exposure.

B: AP view of the ankle using same technique as radiograph A. No change in contrast. Motion during exposure.

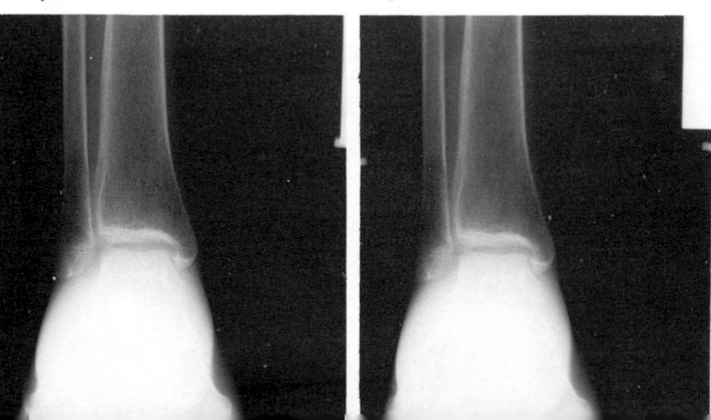

Figure 15-5.

SECTION A. **SECTION B.**

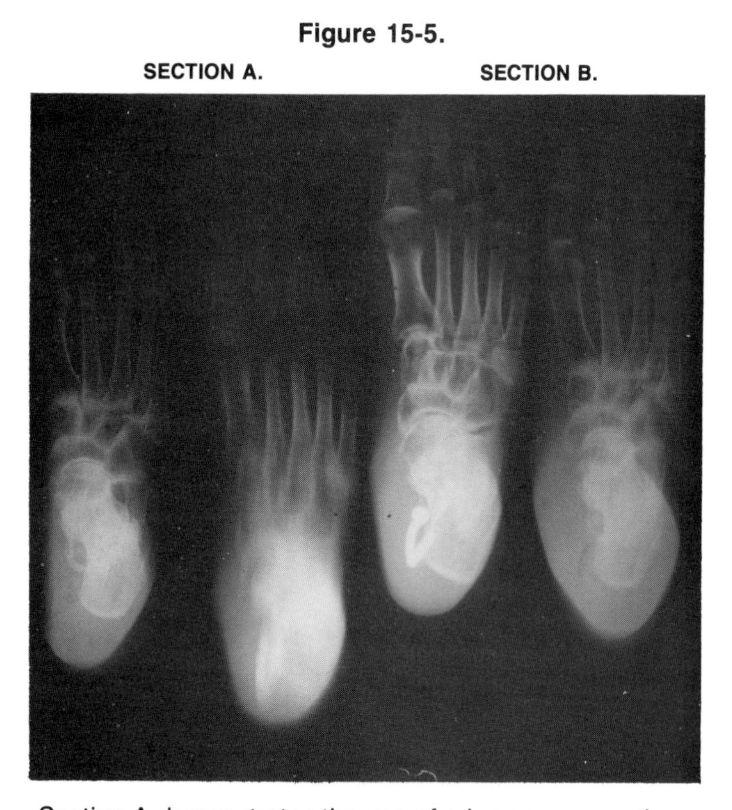

Section A demonstrates the use of a long-exposure time. There is motion on the radiograph with a loss of recorded detail. Section B demonstrates the use of a short-exposure time. There is no motion on the radiograph the part has good recorded detail.

any questions about the procedure. By communicating and gaining the patient's confidence, radiographers can reduce motion on radiographs.

Figure 15-5 demonstrates the use of a short exposure time. Two exposures were made of a manikin's feet, one with motion and one without motion. A 14 × 17 film divided lengthwise was used for both exposures. A nail was placed in one of the feet and a string attached to it. The string was then unwound enough so it would reach behind the radiation barrier. The manikin's feet were placed on one-half of the film and an exposure was made using 50 kVp, 50 mA, 1/5 second, 40″ FFD. During the exposure, the radiographer gently pulled on the string attached to the manikin's foot. The result of this exposure is demonstrated on the film marked A. It is easy to see that the structural lines of the foot are blurred; there are no structural lines and the radiograph has no definition.

On the other half of the film another exposure was made, only this time the machine factors were set on 50 kVp, 200 mA, 1/20 second. Again during the exposure, the radiographer gently pulled on the string attached to the manikin's foot. The result of this exposure is demonstrated on the film marked B. This time the structural lines are very well defined because there is no motion on the film. This experiment proves (illustrates) that if the part moves during a shorter exposure time, better definition of structural details will result.

In medical radiography, use of the automatic exposure timer (AET) is a common practice. AET is another radiographic tool for the radiographer to use, and like other tools, there is an appropriate time and place for its use. An AET is a device that uses an ion chamber to measure the quantity of radiation reaching the film. When a certain amount of radiation reaches the film, the exposure is automatically terminated. Before the AET is placed in service in the clinical envionrment, it is usually calibrated by an x-ray service technician. The radiographer has access to several selection buttons for different desired amounts of radiographic density. The radiographer makes the choice of desired density and then places the machine in AET mode. When the radiographer makes the exposure, the exposure will continue until the proper amount of radiation has been transmitted to the x-ray film.

When radiographing uncoopertive patients or patients who are not able to maintain the proper position, radiographers using AET may have to turn the AET mode off. Revert to the conventional technique timing system, and if possible, increase to a higher mA setting. Some patients recovering from surgery cannot maintain the proper position and do the proper breathing techniques without causing motion on the radiograph. Use of AET in these incidents could result in a reduction in patient care quality because of the risk of unnecessary repeat examinations.

Many chest radiographs are taken using the AET system and this sometimes presents problems for the radiographer. Figure 15-6 shows two lateral chest radiographs taken on the same patient seven days after an operation. The patient was instructed to stand in standard position with the side of his chest next to the chest holder, arms above his head, hands grasping his elbows. The patient was then told to take a big breath and hold it. The radiographer noticed that the patient had a tendency to sway back and forth and when the patient performed the breathing instructions, the swaying back and forth increased. The radiographer had the AET mode on and decided to go ahead and make the ex-

Figure 15-6

Lateral view of chest with automatic exposure timer. Lateral view of chest with conventional timer.

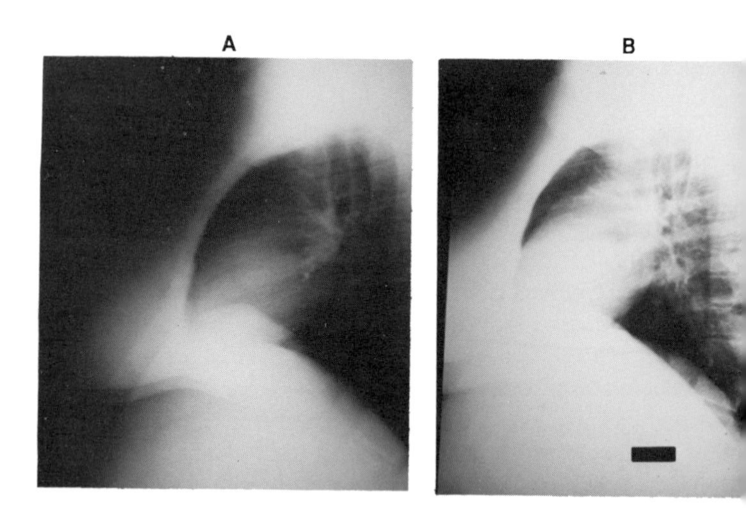

posure. The result of that exposure is represented on Radiograph A. All the structures of the chest are blurred or fuzzy and there is motion of the part caused by the patient's movement. As a result, the radiograph was repeated. The patient was positioned again in the same way; however, this time the radiographer turned the AET mode off. By doing this, the radiographer was in complete control of establishing the necessary exposure time and subsequently, set a very short exposure time on the x-ray equipment. Even though the patient continued to sway back and forth, as was the case in the initial exposure, an acceptable diagnostic radiograph was obtained (Figure 15-6B).

On occasion, the radiographer must make a judgment regarding patients who are uncooperative, unable to maintain the proper position, or unable to perform the proper breathing technique. By using a short exposure time, the radiographer can compensate for the inability of the patient to help in the radiographic process.

Proper immobilization is another way to control accidental motion on a radiography. Radiographers can control the movement of the part during the exposure by using sandbags, sponges, headclamps, cellulose positioning blocks, and tape.

A patient may be very cooperative but is unable to hold motionless under the stress and strain of being injured. By using immobilization, the part will not be able to move during the exposure. The immobilizing device may be placed in position just prior to the exposure and removed immediately after the exposure, so as not to cause any undue discomfort to the patient. Radiographers should formulate a good habit of using the proper immobilization device. Caution should be exercised while working on patients with severe injuries or possible fractures to ensure that pressure is not applied to the injured area.

All equipment should be securely locked in place to prevent movement during the exposure. Movement of the filmholder, x-ray tube, table, or chest holder will also produce a loss of recorded detail on a radiograph.

Filmholders must be secured properly. When working with cross-table laterals, portables, or severe trauma cases, the film may not be placed in an ideal support situation. The radiographer may find that the patient will be unable to support and balance the filmholder during the exposure, causing the filmholder to slip, move, or fall. This, of course, leads to motion on the radiograph. Filmholders can be made more secured by using vertical filmholders, sandbags, immobilization bands, and tape. Making sure that filmholders are properly secured only takes a few moments. This may be considered a better use of time rather than taking repeat radiographs due to unsecured filmholders. Radiograph A in Figure 15-7 shows the result produced on a radiograph when the filmholder is not properly secured.

The x-ray table and table should be locked in place before making an exposure. Some x-ray tables have a floating tabletop, and if it is not locked in the proper place, it may move during the exposure causing a loss of recorded detail on the radiograph. Radiograph B in Figure 15-7 demonstrates the movement of the tabletop during an exposure. Likewise, movement of the x-ray tube during an exposure will cause motion on a radiograph. The radiographer should check to ensure that all locks are in the on

Figure 15-7.

Motion caused by floating tabletop.	Motion caused by filmholder movement.	Motion caused by x-ray tube movement.

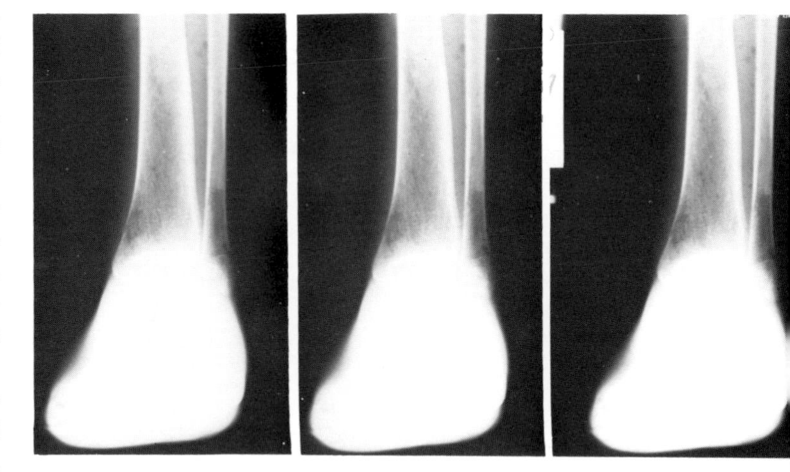

position before making an exposure. If a lock has failed, then one should use sandbags or other devices to ensure there will be no movement during the exposure. Radiograph C in Figure 15-7 shows the result produced when there is movement of the x-ray tube during an exposure.

Size Distortion

Motion has no effect on size distortion (magnification). Movement during an exposure does not change the geometrics of the radiograph. It is true when a part increases in size (magnified) the structure lines become blurred and fuzzy, but this is due to an increase of penumbra and not motion.

Shape Distortion

Motion has no effect on shape distortion. Movement during an exposure will not produce distortion. Shape distortion is produced when the x-ray tube, filmholder, or part being radiographed is angled.

SUMMARY

Table 15-1 shows the different effects that motion has on radiographic quality. Scanning the table quickly, one can see that motion affects only the recorded detail on the radiograph. An increase of motion will decrease the recorded detail, and decreasing the motion will increase the recorded detail on a radiograph. There are several ways in which a radiographer can produce a radiograph free from motion. The best way to eliminate motion is to gain the patient's confidence and cooperation. Effective communication between the patient and radiographer can help the patient understand exactly what is expected during the radiographic procedure. Another way to reduce motion is to use short exposure time and immobilization of the part. Securing the equipment such as, cassette, Bucky tray, or tabletop will keep the equipment from moving during the exposure. It takes only a few moments to ensure that those items have been checked, but the end result may be a radiograph with the best recorded detail.

TABLE 15-1. **Summary of the Effects of Motion on Radiographic Quality.**

IMAGE VISIBILITY	
Radiographic Density	0
Radiographic Contrast	0
IMAGE FORM	
Recorded Detail	–
Size Distortion	0
Shape Distortion	0

\+ = Direct Relationship
– = Inverse Relationship
0 = No Effect

STUDY QUESTIONS

1. List the two types of motion that will produce poor recorded detail on a radiograph.
2. List three ways a radiographer can eliminate motion on a radiograph.
3. Explain how motion can be a useful tool in medical radiography and give examples of procedures in which motion is used to an advantage in radiography.
4. Explain how motion affects radiographic density and contrast.
5. Explain how motion affects the recorded detail and distortion of a radiograph.

BIBLIOGRAPHY

1. Bushong, Stewart. *Radiologic Science for Technologists*. 2nd ed., St. Louis: C. V. Mosby Co., 1980.
2. Cahoon, John B. *Formulating X-Ray Techniques*. 6th ed. Durham, N.C.: Duke University Press, 1974.
3. Curry, Thomas S., James E. Dowdeg, and Robert C. Murry, Jr. *Christensen's Introduction to the Physics of Diagnostic Radiology*. 3rd ed. Philadelphia: Lea & Febiger, 1984.
4. Fuchs, Arthur W. *Principle of Radiographic Exposure & Processing*. Springfield, Ill.: Charles C. Thomas, 1973.
5. Selman, Joseph. *The Fundamentals of X-Ray Radium Physics*. 6th ed. Springfield, Ill: Charles C. Thomas, 1979.

Chapter 16

The Effects of Film Processing on Radiographic Quality

Chapter Outline

1. Learning Objectives
2. Introduction
3. Processing Considerations
4. The Effects of Processing on Image Visibility
   A. Radiographic Density
   B. Radiographic Contrast
5. The Effects of Processing on Image Formation
6. Summary
7. Study Questions
8. Bibliography

Learning Objectives

Upon completion of this chapter, the student should be able to:

1. Name two types of filters that are commonly used in the x-ray darkroom.
2. List five problems that are related to the use of safelights.
3. Explain some of the major differences between manual and automatic processing.
4. Explain how safelights fog and film processing affect the following:
   a. Radiographic density
   b. Radiographic contrast
   c. Recorded detail

Introduction

In most situations the darkroom is where the radiographic process begins and ends. Simply stated, x-ray film is received into the darkroom and then is loaded into filmholders. The filmholder is used in radiography of a patient and is returned to the darkroom for processing. The film is removed from the filmholder and placed into a chemical solution—developer— that will make the latent (invisible) image on the film visible. Another chemical—fixer—will be used to make this image clearer and preserve it permanently. Any mishap along the different steps of film processing could render the film useless. All of the care and time the radiographer invested in the correct positioning and establishing of technique can be destroyed quickly in the darkroom. Because processing of radiographs is a very important step in the radiographic process. individuals working in the darkroom must take as much pride in their work as those working in the radiographic exposure room. This chapter will cover how safelights, handling of film, and processing may affect the quality of the finished radiograph.

Processing Considerations

Safelights

Certain lighting conditions are necessary to prevent fog on the film. Remember fog is an unwanted density on the radiograph, and it can be caused by scatter radiation or light reaching the film. To ensure against darkroom fog production, darkrooms require the use of safelights. Safelights are fixtures that are mounted on the ceiling, wall, or on the processor. Inside the fixture is a white light bulb and over the face of the fixture is a colored piece of glass or plastic called a filter.

Recalling from Chapter 14, radiographic film is sensitive to different colors. For example, a facility that is using calcium tungstate screens would use radiographic film sensitive to blue-violet color. If lanathide screens are used, the radiographic film would be sensitive to blue-green color. A safelight must be compatible with the color sensitivity of the radiographic film. An amber filter, for example, is compatible with a film sensitive to blue-violet and a red filter for film sensitive to blue-green color. Filters can be purchased that are compatible with both types of films and these filters are now widely used.

Just as the name implies, a safelight, *when used properly*, indicates that one can handle the x-ray film without producing an unwanted density or fog on the image. Safelights are not flawless, and it should never be taken for granted that the safelight will always work. There are several potential problems one should be aware of when using safelights. These potential areas of concern involve both human and technical factors. Technical factors, for example, include safelights that may have cracks and have faded. Human factors include wrong wattage of the bulb, and excessive handling of the radiographic film under safelight conditions.

A crack in the filter of the safelight will not usually be visible to the human eye, but x-ray film is sensitive to light and will detect it. This flaw could cause a pattern or a darker density on the film.

Safelights that are operational for twenty-four hours a day may fade faster and transmit a light slightly different from a new safelight. This fading of the filter will permit damaging light to be transmitted through the safelight, and will cause a fogging effect on the radiograph. Safelights should be turned off when not in use in order to prolong the useful life of the filter.

One of the most common faults when using a safelight is using the wrong replacement bulb. Most filters require a bulb of no greater than 15 watts. Some may require 25-watt bulbs, but these are usually used with a fixture that is directed toward the ceiling to provide for indirect lighting. The speed of the film will also dictate the bulb wattage. A replacement bulb of a greater wattage than recommended will create a film with an increased amount of density. Remember, a fast speed film means the film is more sensitive to light, therefore, the faster the speed of the film the less light it requires to expose the film. Consequently, the bulb wattage may have to be decreased with the use of a very fast speed film.

Other problems exist such as, mounting of safelights directly over the feed tray of the processor or over the loading benches, which would cause fog on the finished radiographs.

Remember, x-ray film is eight times more sensitive after exposure, so the film should be removed from the cassette and processed as quickly as possible. Film should not be stacked on top of one another when using a safelight that is mounted as direct lighting or the edge of one film may become imaged on another film. The less time the film is exposed to the safelight, the less chance fog will be produced. With the newer and faster films, it may require a bulb of less than fifteen watts. The exact recommendations of bulb wattage are usually included with the specific film.

Film Processing

Manual processing has been around since the beginning of x-ray technology, but it is fast becoming a lost art due to the arrival of automatic processors in the 1950s. Since that time, the movement from manual to automatic processing has been tremendous. Automatic processing has taken all the steps that were accomplished by an individual and mechanized them. Both types of processing accomplish the same task and require the use of a darkroom, with the exception of some specialized automatic systems. Each method has its own advantages and disadvantages.

One of the advantages of manual processing is the opportunity to see how the latent image becomes a visible image. Students can manually process a radiograph under safelight conditions and see the different changes that take place on the film at the different stages of the processing system. Another advantage is that various types of film can be processed. The thickness of the film emulsion does not make any difference and the darkroom technologist determines how long the film will be left in the developer, fixer, wash, and dryer. If the film has a thick emulsion, then the film may require a longer developing time as compared to a thinner emulsion film which would require a shorter developing time.

When processing films manually, the technologist is in visual contact with all the chemicals and can see if the developer needs to be replenished, or if there is proper flow of water. If the water flow isn't adequate, the film could be left in the wash area for a longer period of time to compensate for the inadequate water flow.

One of the disadvantages of manual processing is there is more of a chance of error. With an individual manually moving the film from tank to tank, there is always the possibility of contamination of the chemicals. The darkroom technologist must maintain the proper chemical concentrations, levels, and temperature. There is an increased amount of film handling and this increases the chance of mishandling which may cause artifacts. It should be remembered that an increase in the handling time under safelight conditions increases the possibilty of fog on the film.

Another area of concern is that the technologist is working constantly with water and electricity. Consequently, there is greater danger of being electrocuted, and the darkroom technologist must practice safe working habits to prevent this from occurring. With manual processing, the darkroom in general is hard to keep clean because of the large amounts of liquid chemicals involved.

The biggest disadvantage of manual processing, however, is the lack of speed with which the films are processed. Approximately ninety minutes are required to move films from the developer tank through the drying cabinet. This limits the efficiency of a radiology department and the consistency of radiographic quality decreases. Efficiency and consistency decrease because of the human element. It requires more time for the darkroom technologist to hang the film on the hanger and manually move the film from tank to tank. What happens if the technologist leaves the film in the developer for a longer or shorter time than is recommended? The optimum quality of the processing of the film will decrease. The radiographic quality will be inconsistent.

Automatic processing units were developed in the 1940s and 1950s, but they came of age in the 1960s. Since 1960 automatic processing has been the choice for processing radiographic films. Automatic processing mechanically incorporates all the steps that are done by manual processing. The automatic processor maintains the length of time the film will be in the developer, fixer, wash, and dryer. It can also regulate the temperature of the different chemicals involved in the processing system, and can control the amount of chemicals that are needed for replenishing.

Just like manual processing, the automatic processor has advantages and disadvantages. With automatic processing, there is less error in processing the film because the human element is virtually eliminated. The film is automatically transported from chemical to chemical in a light-tight machine. The chances of the film being fogged are decreased due to the shorter time of film handling and exposure to safelights. Also the decrease in film handling decreases the amount of artifacts on the radiograph. The chances of the darkroom technologist receiving a severe electrical shock are lessened. The automatic processor does use water and chemicals, but the technologists are not constantly working with their hands in liquids. The processor is grounded so personnel using the equipment are safe from electrical shocks.

Most automatic processors can process a film in ninety seconds as compared to ninety minutes with the manual method. This decrease in processing time increases the work capacity of the department, and patients do not have long extended waiting times. Processing quality also increases with the automatic system, given that the machine maintains the proper levels, concentrations, and temperatures of the chemicals. Radiographs will be processed with the same consistency in degree of preferred contrast.

Automatic processors do have their limitations. One of the things an automatic processor cannot do is mix its own chemicals. The darkroom technologist must be responsible for making sure that the replenishing chemicals are properly mixed, in the proper container, and always available.

The biggest disadvantage to the automatic system is that it is not capable of processing all types of radiographic film. Film that has a thicker emulsion such as, industrial-nonscreen film, cannot be processed automatically. This type of film will not readily transport through the narrow roller system. Likewise the thicker emulsion requires a longer developing time. Since the automatic processor controls how long the film will be in the developer, there is no practical way to process the film for longer periods.

Figure 16-1. H & D Curves. The below curves demonstrate the relationship between the developer temperature and the film density and contrast. As the temperature increases the density increases, and contrast is increased.

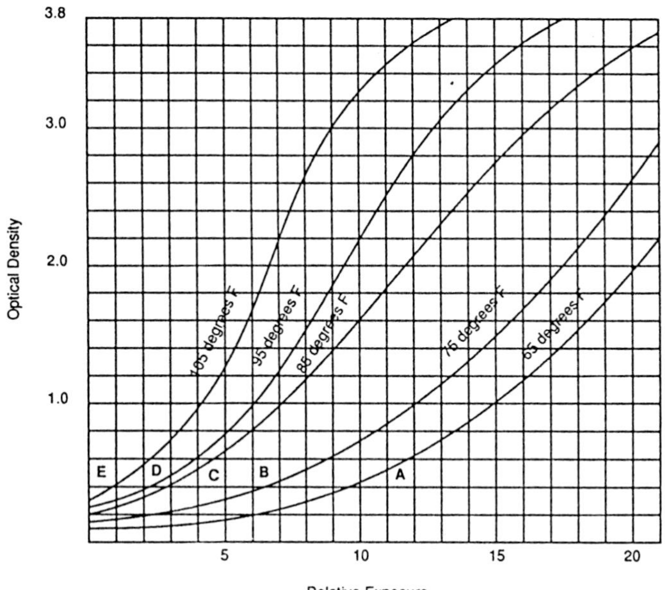

The Effects of Processing on Image Visibility

Radiographic Density

As previously stated, if safelights are not used properly, there will be an increased amount of density in the form of fog on the processed radiograph. The safelight must have the proper bulb wattage, must be mounted correctly, and must have the correct type of unfaded filter for the film type being used. Any one of these factors can cause a fogging effect or an increase in density on the film.

Time and temperature are two important factors when processing a film. In manual processing these two factors go hand in hand. If the temperature of the developer is high, for example, then the film is left in the developer a shorter time; if the developer temperature is too low, the film is left in the developer a longer time. When looking at the final radiographs, there should be no observable differences.

In the automatic system one does not have control over how long the film will remain in each chemical. Consequently, the temperature of the developer, fixer, and dryer must be optimum before processing films. But what happens if the developer is too cold or too hot?

Since the darkroom technologist has no control on how long the film will remain in the developer, the developer temperature should be correctly maintained at all times. This is accomplished by a flow of water to the automatic processor. This water will help the processor cool the developer if it becomes too hot. The developer recirculation system of the automatic processor also has a heat exchanger that will warm the developer if the temperature is too cold.

Figure 16-1 demonstrates the relationship between the developer temperature and the amount of density on a radiograph. Five sensitometric strips were produced and processed at different developer temperatures using an automatic processor. The results of these strips were plotted on a piece of graph paper in the form of H & D curves. Curves A, B, C, D, and E represent the developer temperatures of 65, 75, 85, 95, 105 degrees respectively. Curve A has different densities ranging from .15 to 1.55 as compared to curve C with densities ranging from .18 to 3.07. As the temperature is increased, the amount of density is increased. Curve D represents the proper developer temperature for this particular automatic processor. The densities range from .19 to 3.48, which will give a wide range of density.

The developer temperature on the automatic processor is very important because this chemical controls the amount of radiographic density on the finished radiograph. This temperature must be checked periodically throughout the day to ensure that it is being properly maintained. If not, radiographs that have been properly exposed could be rendered useless because the proper amount of density was not produced on the radiograph.

Radiographic Contrast

Generally speaking, defective safelights will affect the contrast of the radiograph in the middle shades of gray. A review of Chapter 2 on H & D curves is advisable. On an H & D curve the body portion of the curve will be affected by fog. This area with the increase of fog will decrease contrast. In general, the contrast in the toe and shoulder portions will not be affected by fog.

When a radiograph is placed in the developer tank, several reactions take place. First, the activator softens the emulsion so that the reducers are able to get into the exposed silver bromide crystals. There are two reducers in the developer: one will give the film density and the other will give the film contrast. We already know what happens to the density with a change of temperature, but let's look at what happens to the contrast.

Hydroquinone is the reducer that brings out the black tones or different shades of grays on the radiograph. This will also impact on contrast. If the temperature of the developer is below 60 degrees, the hydroquinone will cease its activity. The latent image will be made visible, but the radiograph will lack black tones/shades of gray and contrast will *decrease*. In other words, the radiograph would have no contrast because the radiograph would exhibit very few different shades of grays. If the temperature of the developer is increased significantly higher than normal, the contrast will also *decrease*. This is the reason why the developer temperature is so important in the processing system.

Figure 16-2 is a series of radiographs of an anterior-posterior view of the skull. All of these radiographs were obtained using the same exposure factors. The only difference in producing these radiographs was the temperature under which they were developed. The radiograph on the far left used a developer temperature of 65 degrees while the radiograph on the far right used 95 degrees. As the developer temperature increased, the contrast on the radiograph increased.

Recorded Detail

Recorded detail of a radiograph implies sharpness of the image structural lines of the image details. Film processing and safelights will not affect this quality factor, except in circumstances where the contrast and density have been significantly altered and the structure film cannot be "seen."

Size Distortion

Size distortion/magnification relates to the enlargement of the part being radiographied. Processing has no effect on size distortion.

Shape Distortion

Shape distortion is affected by the misalignment of the film, part, or x-ray tube. Consequently, processing has no effect on shape distortion.

SUMMARY

Table 16-1 summarizes the different relationships that processing has on radiographic quality. A misuse of safelights can cause an extra density or fog on the finished radiograph. This extra density may alter the amount of contrast, which may require the radiograph to be repeated. Remember safelights can be faulty because of filters being cracked or faded. Other problems include improper installation (direct vs. indirect) and bulb wattage in the safelight.

Film processing plays an important part on the amount of density and contrast a radiograph will possess. Temperature is the critical factor in both manual and automatic processing that will determine the optimum amount of density and contrast. As the developing temperature increases the amount of density increases and conversely, as the developer temperature decreases, the amount of density decreases. Contrast can also be altered by a change of temperature. Normally, this will not be a factor unless the temperature of the developer is significantly altered from the normal. If the temperature is too cold or hot, contrast will *decrease* in both situations. There will be a lack of differences in the shades of gray and the radiograph will appear to have the same shade of gray throughout the radiographic image.

Figure 16-2.

AP skull developed at a temperature of 65 degrees. AP skull developed at a temperature of 75 degrees. AP skull developed at a temperature of 85 degrees. AP skull developed at a temperature of 95 degrees.

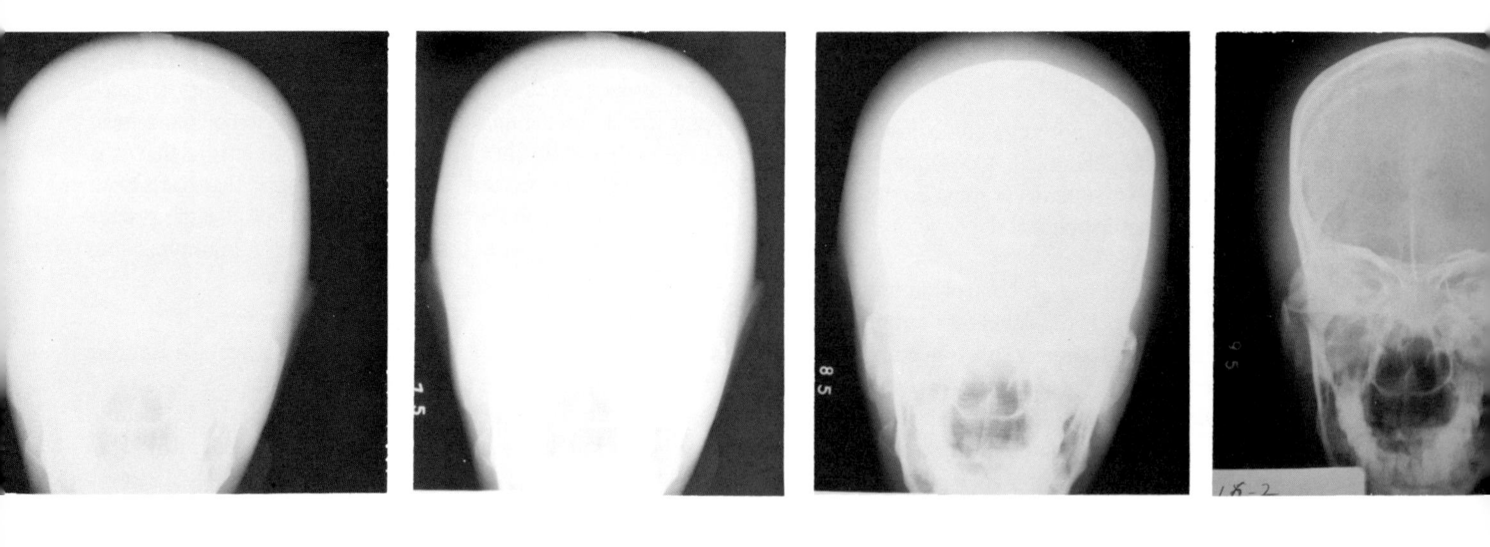

TABLE 16-1. **Effects of Processing on Radiographic Quality.**

IMAGE VISIBILITY	
Radiographic Density	
Safelight Fog	+
Film Processing	+
Radiographic Contrast	
Safelight Fog	–
Film Processing	– & +
IMAGE FORM	
Recorded Details	0
Size Distortion	0
Shape Distortion	0

+ = Direct Relationship
– = Inverse Relationship
0 = No Effect

STUDY QUESTIONS

1. Explain the different types of filters that are needed/used in radiology.
2. List three problems associated with safelights.
3. Explain one advantage to manual processing.
4. Explain one disadvantage of automatic processing.
5. Explain how film processing affects density.
6. Explain how safelight fog affects density and/or contrast.
7. What is the relationship between the developer temperature and the amount of density and contrast on a radiograph. Explain your answer.

BIBLIOGRAPHY

1. Curry, Thomas S., James E. Dowdey, and Robert C. Murray, Jr. *Christensen's Introduction to the Physics of Diagnostic Radiology*. 3rd ed. Philadelphia, Penn.: Leas & Febiger, 1984.
2. Fuchs, Arthur W. *Principle of Radiographic Exposure & Processing*. Springfield, Ill: Charles C. Thomas, 1973.
3. *Safelight in the Automated Radiographic Darkroom*. Rochester: Eastman Kodak Company, 1978.
5. *The Fundamentals of Radiography*. Rochester: Eastman Kodak Company, 1980.

Chapter 17

Radiographic Exposure Conversion Problems

Chapter Outline

1. Introduction
2. Learning Objectives
3. Milliamperage-Second (mAs) Conversions
4. Intensification Screen/mAs Conversion
5. Focal Film Distance/mAs Conversion
6. kVp/mAs Conversions
7. Grid Conversions
8. Multiple Conversions
9. Radiographic Techniques Affecting Density
10. Study Questions
11. Bibliography

Learning Objectives

Upon completion of this chapter, the student should be ble to:

- Calculate mA, exposure time, and mAs values using the mAs conversion formula.
- Determine mAs or Screen Speed values when converting from one screen speed to another using the mAs/intensifying screen conversion table.
- Calculate screen speed and mAs values using the intensification factor/mAs formula.
- Calculate radiation intensity or distance values using the inverse square law formula.
- Calculate mAs and distance values that equalize radiographic densities using the Focal Film Distance/mAs conversion formula.
- Calculate mAs and kVp values that equalize radiographic densities using the 15 percent rule.
- Calculate mAs or kVp values that equalize radiographic densities using grid conversion factors based on a given set of grid ratios.
- Calculate mA, exposure time, kVp, grid ratio, screen speed, or FFD using strategies for the solution of multiple conversions that equalize radiographic densities.
- Determine the one exposure technique with the greatest or least amount of radiographic density based on a series of similar techniques.

Introduction

Sooner or later all radiographers are placed in situations in which they will be required to convert radiographic exposure and imaging variables for atypical radiographic procedures, such as portable examinations. When performing even the simplest types of radiographic procedures the exposure and imaging variables such as focal film distances, grids and screens, may occasionally require deviation from the normal departmental standards. Effective and efficient response to such deviation requires clear thinking and the ability on the part of the radiographer to adjust the exposure technique accordingly so that the quality of the eventual radiograph produced is not affected. For the sake of maintaining an optimum level of quality patient care, such exposure and imaging adjustments must not be done thoughtlessly by guessing the compensating exposure and imaging factors. The radiographer must possess the analytical tools necessary to calculate compensations that are required when deviations from normal radiographic technique values are encountered.

After obtaining the solution to any radiographic exposure and imaging problem, the reader is encouraged to reflect for a moment and try to determine if the direction and magnitude of the value of the answer arrived at coincides with what might be theoretically expected. This way, one may isolate situations in which multiplication should have been used instead of division, or vice versa. Likewise, if the magnitude of the value seems to coincide with knowledge about the inverse square law (or any other applicable radiographic theory) then the correct answer was probably obtained. However, if a magnitude, for example, of approximately four times more than the original mAs value was expected and the answer reflects four times less, something went wrong in the calculation. These are indications that calculations need to be rechecked.

The reader is encouraged not to take issue with the validity of a given radiographic exposure condition from the problems presented. The purpose of this chapter is to learn solution strategies that impact on maintaining radiographic quality. Regardless of the many exposure variables that may be used, if the appropriate strategy is learned, a degree of confidence for solving these types of problems will be realized. Thus, the learner will be more apt to tackle and solve radiographic exposure and imaging conversion problems on a more objective basis rather than shying away from them through subjective problem solving techniques or guessing. In this manner, objective solution strategies will contribute towards a greater degree of radiographic patient care quality.

Milliamperage-Second (mAs) Conversions

The easiest calculation learned by radiographers is the milliamperage-second conversion formula. This formula is employed when a new mAs, mA, or exposure time is desired for a given radiographic situation. Simply, the mAs formula states that milliamperage multiplied by exposure time will yield milliampere-seconds, or mAs.

mAs Formula

Milliamperage (mA) x Seconds (s) = Milliamperage-Seconds (mAs)

Problem # 1

Given 1/10th of an exposure time setting at 100 milliamperes, calculate the correct milliamperage-seconds (mAs).

Solution

100 mA x 1/10 s = 10 mAs

Problem # 2

Given 10 mAs, calculate the correct exposure time setting using 100 mA.

Solution

$$100 \text{ mA} \times ?\ \text{s} = 10 \text{ mAs}$$
$$\text{s} = \frac{10\text{mAs}}{100 \text{ mA}}$$
$$\text{s} = 1/10 \text{ or } .10$$

Problem # 3

Given 10 mAs, calculate the correct mA setting if the exposure time is 1/10th of a second.

Solution

$$?\ \text{mA} \times 1/10 \text{ s} = 10 \text{ mAs}$$
$$\text{mA} = \frac{10\text{mAs}}{.10\text{s}}$$
$$\text{mA} = 100$$

Intensifying Screen/mAs Conversions

The manufacturers of intensifying screens place a relative speed value on their screens based on screen type, film used, and exposure settings. The intensification factor of a screen is determined by the ratio of the exposure value without screens (direct exposure) divided by the exposure required with screens to provide the same photographic effect.

$$\text{Intensification Factor} = \frac{\text{Exposure (mAs) Without Screens}}{\text{Exposure (mAs) With Screens}}$$

When the intensification factors are known for particular screen types, screen conversion factors may be established based on the ratio of known intensification factors for different screens. This allows the radiographer to adjust exposure factors proportionately depending on which type of intensifying screen is used. See Table 17-1.

$$\text{Screen Conversion Factor} = \frac{\textit{Intensification Factor \#1}}{\text{Intensification Factor \#2}}$$

TABLE 17-1. **Screen Conversion Factors Based on Known Screen Intensification Factors Used to Determine New mAs Values (÷ = divide by, x = multiply by)**

Known IF	IF	20 CBH	2 SLOW	1 PAR	0.50 HIGH	0.35 SUPER FAST
20	CBH	**1**	**÷10**	**÷20**	**÷40**	**÷57**
2	SLOW	**10x**	**1**	**÷2**	**÷4**	**÷5.7**
1	PAR	**20x**	**2x**	**1**	**÷2**	**÷2.9**
0.50	HIGH	**40x**	**4x**	**2x**	**1**	**÷1.4**
0.35	SUPER FAST	**57x**	**5.7x**	**2.9x**	**1.4x**	**1**

(Directions: Convert from left "IF" column to top "IF' column. For example, when converting from Slow Speed Screens (IF = 2) at left to High Speed Screens (IF = .50) at top will require division of the original mAs factor by 4 so that equivalent radiographic densities are maintained.)

Problem # 1

What would be the conversion factor for a new mAs value when converting from par-speed screens to high-speed screens using Table 17-1?

Solution

Starting from the left-hand column of Table 17-1 find par-speed screens. Follow horizontally until high-speed screens are intersected from above. At the intersect is the screen conversion factor which indicates that when converting from par- to high-speed screens the radiographer would divide the original mAs in half. For example, if 50 mAs was used using par-speed screens, then 25 mAs would be required to maintain the same photographic effect.

Problem # 2

A radiographer is required to use a technique employing 40 mAs with high-speed screens. What would his new exposure technique have to be if he were to switch and use slow-speed screens instead?

Solution

Try the ''intensification factor/mAs conversion formula'' for solving this problem.

xample:

$$\frac{\text{Old IF}}{\text{New IF}} = \frac{\text{Old mAs}}{\text{New mAs}}$$

$$\frac{0.50}{2.00} = \frac{40 \text{ mAs}}{? \text{ mAs}}$$

Cross Multiply

$$0.50 \text{ IF } (? \text{ mAs}) = 2.00 \text{ IF } (40 \text{ mAs})$$

$$? \text{ mAs} = \frac{2.00 \text{ IF } (40 \text{ mAs})}{0.50 \text{ IF}}$$

$$? \text{ mAs} = 160$$

The above technique for solving screen conversion problems esulted in the same answer for a new mAs value that would have een obtained form using the screen conversion factors in Table 7.1. Application of this intensification factor/mAs conversion ormula may be easier than memorizing or having to refer to creen conversion factors from Table 17-1.

Focal Film Distance/mAs Conversions

A frequent problem that the radiographer will encounter is to djust the exposure (mAs) due to a change in the distance from e x-ray tube to the film. Remember from the inverse square w that the radiation intensity is inversely proportional to the quare of the distance. That is, if the intensity is 10 Roentgens t 40 inches from the x-ray tube, then it will measure 40 Roentens at half this distance, 20 inches from the x-ray tube.

alculation for the Inverse Square Law

$$\frac{\text{Original Intensity}}{\text{New Intensity}} = \frac{\text{New Focal Film Distance}^2}{\text{Original Focal Film Distance}^2}$$

$$\frac{10 \text{ Roentgens}}{? \text{ Roentgens}} = \frac{20 \text{ Inches}^2 \ (400 \text{ Inches})}{40 \text{ Inches}^2 \ (1600 \text{ Inches})}$$

Cross Multiply

$$? \text{ Roentgens } (400 \text{ Inches}) = 10 \text{ Roentgens } (1600 \text{ Inches})$$

$$? \text{ Roentgens} = \frac{10 \text{ Roentgens } (1600 \text{ Inches})}{(400 \text{ Inches})}$$

$$\text{Roentgens} = 40$$

Because of the inverse square law, the radiographer has a tool or determining the appropriate mAs value that would be needed maintain equal radiographic densities if focal film distance is aried. Focal film distances are often varied during emergency, auma, and surgical radiography, and other situations in which e portable or stationary x-ray machine is used. Correct comensations in exposure must be maintained if equivalent adiographic densities are to be maintained. (Remember the eciprocity law?)

Students often confuse the proper application of the inverse quare law in calculating new mAs values when changes from e original focal film distance are experienced. In solving the verse square law problem radiation intensity levels at different istances from the radiation source are being determined. The adiographer is not so much interested in the value of the intensity as it varies at different distances but the amount of exposure intensity (mAs) needed to maintain the original intensity level (and resultant radiographic density) at different distances. (See Figure 17-1.)

A different but smaller calculation is required for determining the appropriate mAs value to use as distance changes in order to maintain equivalent densities. In calculating for a new mAs as FFD changes, the original mAs is proportional to the square of the distance as is the case when calculating intensity levels from the inverse square law. (Again see Figure 17-1.) Always remember to square the distance.

Problem # 1

A radiographic procedure requires an exposure setting of 40 mAs at a 40-inch FFD. However, due to a restriction, only a 20-inch FFD may be used. What is the new mAs value that would maintain an equivalent radiographic density?

Figure 17-1. Demonstration of the relationship between the application of the Inverse Square Law and corresponding exposure (mAs) values that result in equivalent radiograph densities as FFD changes.

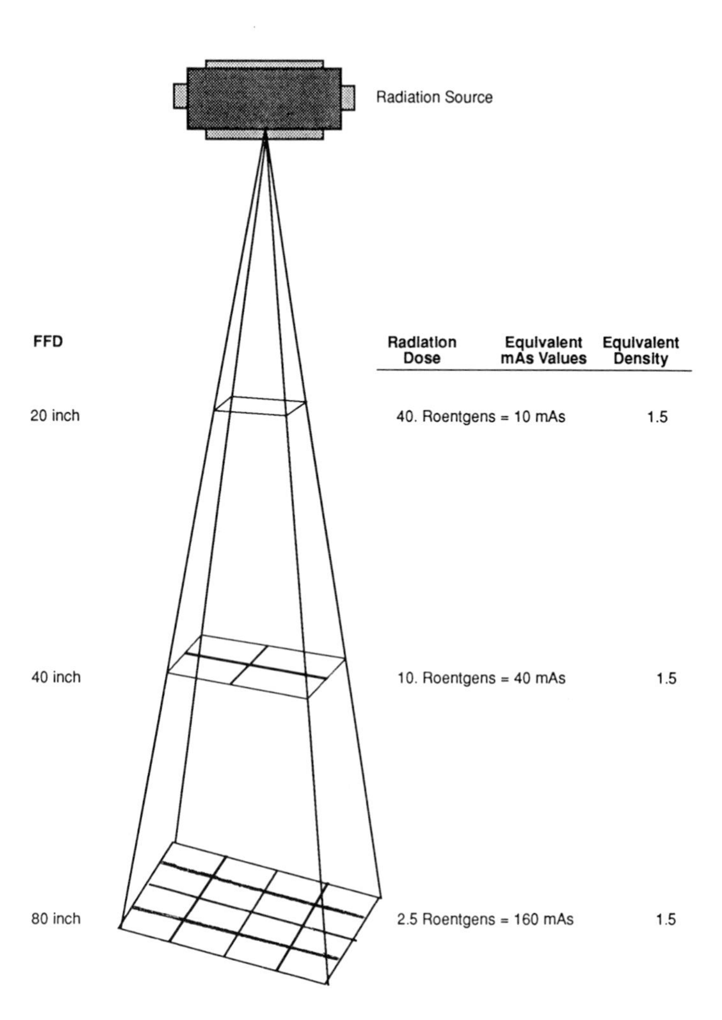

Solution

$$\frac{\text{Original mAs}}{\text{? New mAs}} = \frac{\text{Original FFD}^2}{\text{New FFD}^2}$$

$$\frac{\text{40 mAs}}{\text{? mAs}} = \frac{\text{40 Inches}^2\ \text{(1600 Inches)}}{\text{20 Inches}^2\ \text{(400 Inches)}}$$

Cross Multiply

$$\text{? mAs (1600 Inches)} = \text{40 mAs (400 Inches)}$$

$$\text{? mAs} = \frac{\text{40 mAs (400 Inches)}}{\text{(1600 Inches)}}$$

$$\text{New mAs} = 10$$

Problem # 2

A radiographic procedure requires an exposure setting of 40 mAs at a 40-inch FFD. Due to a restriction, however, only an 80-inch FFD may be used. What is the new mAs value that would maintain an equivalent radiographic density?

Solution

$$\frac{\text{40 mAs}}{\text{? mAs}} = \frac{\text{40 Inches}^2\ \text{(1600 Inches)}}{\text{80 Inches}^2\ \text{(6400 Inches)}}$$

Cross Multiply

$$\text{? mAs (1600 Inches)} = \text{40 mAs (6400 Inches)}$$

$$\text{? mAs} = \frac{\text{40 mAs (6400 Inches)}}{\text{(1600 Inches)}}$$

$$\text{? mAs} = 160$$

Problem # 3

(This problem should test for mastery of the application of mA/Focal Film Distance conversions.) A unique radiographic procedure requires exposure values of 160 mAs at an 80-inch FFD. What would the new FFD be to maintain equivalent radiographic densities if 10 mAs is used?

Solution

$$\frac{\text{160 mAs}}{\text{10 mAs}} = \frac{\text{80 Inches}^2\ \text{(6400 Inches)}}{\text{? Inches}^2}$$

Cross Multiply

$$\text{? Inches}^2\ \text{(160 mAs)} = \text{10 mAs (6400 Inches)}$$

$$\text{? Inches}^2 = \frac{\text{10 mAs (6400 Inches)}}{\text{160 mAs}}$$

$$\text{? Inches}^2 = 400$$

$$\text{? Inches} = \sqrt{400}$$

$$\text{? Inches} = 20$$

kVp/mAs Conversions

From studying radiation physics radiographers learn that a 15 percent increase or decrease in kVp affects radiographic density as if mAs were respectfully doubled or halved. This is generally termed the "15 Percent Rule." (A simplified version of this rule is the Rule of 10, see page 44.) A 15 percent increase of the original kV does not double the x-rays emitted from an x-ray tube but is equivalent to doubling the mAs to obtain a given film blackening effect on the radiograph. In actuality, the kVp would have to be raised by approximately 40 percent if double the amount of x-rays were desired; however, only 15 percent is needed for a doubling of radiographic density since the increase in penetration results in more radiation ionizing silver bromide crystals in the x-ray film emulsion (Bushong, 1980). (A 15 percent increase or decrease at lower kV values, 50 to 80, will be correspondingly lower than changes made at higher kV values, 80 to 140 kV.)

Knowing this information will allow the radiographer to make some wise decisions. Given a choice between doubling the mAs or increasing the original kVp by 15 percent (approximately 10 kV), what would the most logical choice be if twice the amount of radiographic density were desired? In the opinion of the authors, a logical choice would be to increase the kVp by 15 percent. The subject contrast would change towards a longer scale which tends to enhance visibility of image details. At the same time, the patient would receive considerably less radiation exposure than if the mAs were doubled. Likewise, the x-ray tube would experience a reduction in heat units and this would assist in extending x-ray tube life. (The formula for calculating heat units using a three-phase x-ray machine is: Heat Units = mA x exposure time x kV x 1.35. The 1.35 allows for the additional heat generated from three-phase generators compared to single-phase generators.)

For example, on a single-phase generator with original radiographic technique of 10 mAs x 60 kVp = 600 Heat Units which of the following two options for maintaining similar densities results in fewer heat units?

A. Double the Radiographic Density by 2 x Original mAs
   20 mAs x 60 kVp = 1200 Heat Units

B. Double the Radiographic Density by "15 Percent Rule"
   10 mAs x 60 x 15% = 690 Heat Units

If a radiographer wanted to decrease the radiographic density by half its original amount, should the mAs be halved or should there be a reduction of 15 percent in kVp? In the opinion of the authors, a more beneficial choice might be to reduce the original mAs value by one-half. Obviously, from the above example one observes that this would result in the greatest reduction in heat units and corresponding radiation output thereby reducing exposure levels to the patient. With no change in kV, contrast would remain stable with the higher kV levels working to provide greater visibility of image details.

At this point the reader may be confused about the application of the 15 percent rule. Remember, when a decision is made to double radiographic density from an original technique (for example, 80 kV at 25 mAs), if 80 kV were increased 15 percent to 92 kV with a subsequent reduction of mAs by one-half to 12.5 mAs, there would be no significant change in radiographic density. There would, however, be an alteration towards longer scale subject contrast since a higher kV is now being used. (Likewise, a 15 percent decrease from 80 kV to 68 kV with a compensating doubling of 25 mAs to 50 mAs would maintain equal radiographic densities and alter subject contrast towards a shorter scale.) Notice that a doubling of radiographic density in one direction from 80 kV to 92 kV, was offset by halving radiographic density in the other direction, from 25 mAs to 12.5 mAs. Consequently, no significant radiographic density change took place.

In order to make a radiographic density change from the original radiographic technique, either kV or mAs should be changed. Not both! Therefore, if twice or half the radiographic density is desired to compensate for a bad first exposure try, the radiographer would be wise in choosing the exposure factor that has the greatest effect for reducing radiation to the patient and operator.

Problem # 1

An original technique of 100 mAs at 80 kVp is being used for a radiographic projection. A second exposure is required to double the radiographic density without increasing radiation dose to the patient. What would be the new radiographic technique?

Solution

Use 15 Percent Rule

80 kVp x 15% = 12 kVp + 80 kVp = 92 kVp

New Radiographic Technique = 100 mAs at 92 kVp

Problem # 2

An original technique of 30 mAs at 70 kVp is being used for a radiographic projection. A second exposure is required to halve the original radiographic density. What would be the new technique if a corresponding decrease in radiation dose to the patient is desired.

Solution

Since halving the original mAs is more beneficial than reducing kVp by 15 percent to achieve a halving effect on dose and radiographic density, a radiographic technique of 15 mAs at 70 kVp should be selected.

Grid Conversions

The importance of using a grid during radiographic procedures involving thicker anatomical part thicknesses or kV levels above 60 kV or 70 kV should not be underestimated. When used properly, a grid will clean up significant amounts of scatter radiation created from the interaction of x-rays with matter. (Table 17-2.)

TABLE 17-2. **Comparison of Scatter Removal According to Grid Ratio.**

Grid Ratio	Percent of Total Scatter Removed
No Grid	–
5:1 Grid	82%
8:1 Grid	90%
12:1 Grid	95.5%
16:1 Grid	96%

(Source: *Characteristics and Applications of X-Ray Grids*, Liebel-Flarsheim, Cincinnati, Ohio, 1983, p. 7.)

When using a grid, the radiographic exposure (mAs or kV) must be increased in order to compensate for the absorption of the primary radiation due to the greater thicknesses of higher grid ratios (Liebel-Flarsheim, 1983). Table 17-3 illustrates grid conversion factors for mAs and kVp values that may be used in radiographic imaging to overcome reduction in radiographic density as a result of grid absorption characteristics.

TABLE 17-3. **Grid Conversion Factors for mAs or kVp Values.**

Grid Type	mAs Factor	or	kVp Factor
Non Grid	= 1 x mAs		0 kVp
5:1	use 2 x mAs		add 8 kVp
6:1	use 3 x mAs		add 12 kVp
8:1	use 4 x mAs		add 20 kVp
10:1	use 5 x mAs		add 20 kVp
12:1	use 5 x mAs		add 23 kVp
16:1	use 6 x mAs		add 25 kVp

(Source: Adapted from Stevens and Phillips, *Comprehensive Review for the Radiologic Technologist*, C. V. Mosby, St. Louis, 1978, p. 121.)

Problem # 1

A nongrid radiographic technique calls for the use of 50 mAs at 75 kVp. What would be the new mAs if a 6:1 grid were used?

Solution

$$\frac{\text{New Grid Conversion mAs Factor}}{\text{Old Grid Conversion mAs Factor}} = \frac{\text{New mAs}}{\text{Old mAs}}$$

$$\frac{3}{1} = \frac{?\ \text{New mAs}}{50\ \text{mAs}}$$

Cross Multiply

$$?\ \text{mAs}\ (1) = 50\ \text{mAs}\ (3)$$

$$?\ \text{mAs} = \frac{50\ \text{mAs}\ (3)}{(1)}$$

$$\text{New mAs} = 150$$

Problem # 2

A 16:1 grid technique calls for the use of 160 mAs at 80 kVp. What would the new mAs be if an 8:1 grid were used?

Solution

$$\frac{4}{6} = \frac{?\ \text{New mAs}}{160\ \text{mAs}}$$

Cross Multiply

$$?\ \text{mAs}\ (6) = 160\ \text{mAs}\ (4)$$

$$?\ \text{mAs} = \frac{160\ \text{mAs}\ (4)}{(6)}$$

$$\text{New mAs} = 106$$

Problem # 3

A nongrid radiographic technique calls for 100 mAs at 70 kVp. What would the new kVp be if a 16:1 grid was used.

Solution

By consulting Table 17-3, add 25 kV if going from nongrid to a 16:1 grid. Therefore, the new kVp value would be 70 kVp + 25 kVp = 95 kVp.

Problem # 4

A radiographic technique uses a 12:1 grid at 200 mAs at 100

kVp. What would be the new kVp if a nongrid technique were used?

Solution

By consulting Table 17-3, if going from a 12:1 grid to nongrid instead of adding 23 kVp, subtract 23 kVp from the original kVp. Therefore the new kVp to use going from a 12:1 grid to nongrid is 100 kVp - 23 kVp = 77 kVp.

Problem # 5

A radiographic technique calls for the use of a 6:1 grid at 300 mAs at 75 kVp. What would be the new kVp value if the 16:1 grid were used?

Solution

This problem appears a little more difficult at first. However, using a little common sense, it is easily solved. If going from a nongrid to 16:1 grid, then the kVp would increase by 23 kVp. At 6:1 kVp the value increase is only 12 kVp. Therefore, subtracting the 6:1 value of 12 kVp from the 16:1 value of 25 kVp results in a 13 kVp difference (25 kVp - 12 kVp = 13 kVp). Since the direction of grid ratio change is to be increased from 6:1 to 16:1, the 13 kVp difference would be added to the original kVp value. In other words, the new kVp at 16:1 would be 75 kVp + 13 kVp = 88 kVp. (A more simplified solution strategy might be to convert the 6:1 technique to nongrid technique. Then, convert nongrid to 16:1 technique.)

Problem # 6

A radiographic technique calls for the use of a 12:1 grid at 10 mAs at 80 kVp. What would the new kVp be if a 5:1 grid were used instead?

Solution

This problem is identical to problem # 5 except for the direction of change in kVp. The difference between kVp values for 12:1 and 5:1 grid ratios is 15 kVp (23 kVp - 8 kVp = 15 kVp). Since the direction of grid ratio change is decresed from a 12:1 to a 5:1 the 15 kVp difference would be subtracted from the original kVp of 80 kVp at the 12:1 grid ratio. Therefore, the new kVp at 5:1 would be 80 kVp - 15 kVp = 65 kVp.

A final word about the practical application of grid conversions will now be given. Remember, when given a choice of changing from kVp or mAs in a previous section titled ''mAs and kVp Conversion,' it was better to choose a change in kV when the exposure needed to be increased. This was done for radiation protective purposes and as a measure of extending tube life. Also remember, when given the choice of decreasing mAs or kV, it is better to reduce mAs since this will result in the greatest reduction of radiation to the patient, extend tube life, and maintain an optimum level of contrast. The same logic might be employed when given the choice between changing kVp or mAs during grid conversions. When increasing exposure try to increase kVp. When decreasing exposure try to decrease mAs.

Mutiple Conversion Problems

Thus far, this chapter has focused on the identification of problems and their solutions in radiographic exposure that affect imaging with respect to mAs, intensifying screens, FFD, kVp, and grids. These are the exposure variables that the radiographer may often encounter that will require appropriate conversion calculations. The real challenge now comes when problems requiring multiple conversions surface. Much practice and careful development of functional and reliable strategies for realizing solutions is demanded if the radiographer aspires to be proficient at dealing with multiple conversion problems. The reader is encouraged to concentrate more on solving the problems rather than the plausability of the radiographic technical factors to a particular radiographic situation.

Problem # 1 (Finding a new exposure time)

A radiographic technique calls for 200 mA at 1/10th of a second, 65 kVp, 12:1 grid, high-speed screens, and a 40-inch FFD. What would a new exposure time be in order to maintain equivalent radiographic densities, if the radiographic technique was changed to 300 mA, 77 kVp, 5:1 grid, slow-speed screens, and 20-inch FFD.

Solution

First, order both sets of radiographic techniques.

Original Technique	*New Technique*	
200 mA	300 mA	(To be used
1/10 sec	? sec	at final
65 kVp	77 kVp	solution)
12:1 grid	5:1 grid	
High-Speed	Slow-Speed	
40-inch FFD	20-inch FFD	

Second, determine mAs of original technique = 20 mAs.
Third, convert 20 mAs at 65 kVp to 77 kVp = 10 mAs.
Fourth, convert 10 mAs at 12:1 grid to 5:1 grid = 4 mAs.
Fifth, convert 4 mAs at high speed to slow speed = 16 mAs.
Sixth, convert 16 mAs at 40-inch FFD to 20-inch FFD = 4 mAs.
Seventh, determine new exposure time from 4 mAs at 300 mA =

$$300 \text{ mA } (? \text{ sec}) = 4 \text{ mAs}$$

$$? \text{ sec} = \frac{4 \text{ mAs}}{300 \text{ mA}}$$

$$? \text{ sec} = .013$$

Problem # 2 (Finding a new mAs)

A radiographic technique calls for 600 mA at ½ seconds, 90 kVp, 12:1 grid, high-speed screens, and a 40-inch FFD. What would the new mAs be in order to maintain equal radiographic densities if the technique were changed to 105 kVp, nongrid, nonscreen; CBH, and 60-inch FFD?

Solution

First, order both sets of radiographic techniques.

Original Technique	*New Technique*
600 mA	? mAs
½ sec	
90 kVp	105 kVp
12:1 grid	nongrid
high-speed	nonscreen
40-inch FFD	60-inch FFD

Second, determine original mAs value = 300 mAs
Third, convert 300 mAs at 90 kVp to 105 kVp = 150 mAs
Fourth, convert 150 mAs at 12:1 grid to nongrid = 30 mAs
Fifth, convert 30 mAs at high speed to nonscreen = 1200 mAs
Sixth, convert 1200 mAs at 40-inch FFD to 60-inch FFD = 2700 mAs (Answer is 2700 mAs). What would the exposure time have to be if 600 mA were used? Answer: 4.5 seconds.

Problem # 3 (Finding a new grid conversion factor)

A radiographic technique calls for 100 mA at 1 second, 65 kVp, nongrid, nonscreen, at 40-inch FFD. What would the new grid ratio have to be if 50 mA at ½ second, at 65 kVp were used?

Solution

First, order both sets of radiographic techniques.

Original Technique	*New Technique*
100 mA	50 mA
1 sec	½ sec
nonscreen	high-speed screen
65 kVp	65 kVp
40-inch FFD	40-inch FFD
nongrid	? grid

Second, determine the original mAs = 100 mAs
Third, determine new mAs = 25 mAs (Remember to set aside and use in the final solution set)
Fourth, convert 100 mAs at nonscreen to HS Screen = 2.5 mAs
Fifth, kVps are the same, no need to convert. Ignore.
Sixth, solve for mAs grid conversion factor and match with closest known grid ratio from Table 17-3.

$$\frac{\text{Original mAs}}{\text{New mAs}} = \frac{\text{Original Grid Conversion mAs Factor}}{\text{New Grid Conversion mAs Factor}}$$

$$\frac{2.5 \text{ mAs}}{25 \text{ mAs}} = \frac{1}{x}$$

Cross Multiply

$$x\,(2.5 \text{ mAs}) = 1\,(25 \text{ mAs})$$

$$x = \frac{1\,(25 \text{ mAs})}{(2.5 \text{ mAs})}$$

$$x = 10$$

The new grid conversion mAs factor for this problem is 10. After consulting Table 17-3 the highest grid conversion mAs factor available is 6 x mAs and this corresponds to using a 16:1 grid; therefore, the new grid ratio to use with the above change in technique is 16:1, since it is the highest available.

Problem # 4 (Finding a new grid conversion factor)

A radiographic technique calls for 600 mA at 1/10 of a second, 80 kVp, and a 12:1 grid. What would the new grid ratio be in order to maintain equal densities if 300 mA at 1/10 of a second at 80 kVp were used instead?

Solution

First, order both sets of radiographic techniques.

Original Technique	*New Technique*
600 mA	300 mA
1/10 sec	1/10 sec
80 kVp	80 kVp
12: 1 grid	? grid

Second, determine mAs value for original technique = 60 mAs
Third, determine mAs value from new techique = 30 mAs (Hold until final solution set).
Fourth, kVps are same, no need to convert. Ignore.
Fifth, solve for grid conversion factor and match with closest known grid ratio from Table 17-3.

$$\frac{\text{Original mAs}}{\text{New mAs}} = \frac{\text{Original Grid Conversion mAs Factor}}{\text{New Grid Conversion mAs Factor}}$$

$$\frac{60 \text{ mAs}}{30 \text{ mAs}} = \frac{5}{x}$$

Cross Multiply

$$x\,(60 \text{ mAs}) = 5\,(30 \text{ mAs})$$

$$x = \frac{5\,(30 \text{ mAs})}{(60 \text{ mAs})}$$

$$x = 2.5 \text{ (New Grid Conversion mAs Factor)}$$

The new grid conversion factor to use with the new radiographic technique in problem # 4 is 2.5. After consulting Table 17-3, one does not find a grid conversion factor of 2.5. Therefore, round off the conversion factor to the next highest value if the decimal is .5 or greater. Likewise, round off to the next lowest value if the decimal is less than .5. In this case, 2.5 would round off to the next highest value of 3.0. After consulting Table 17-3, one can find an mAs grid conversion factor of 3; therefore, a grid conversion factor of 3 corresponds to the use of a 6:1 grid. The new grid of choice for use with the new technique is a 6:1 grid. (Considering that the original mAs of 60 was reduced by half to 30 mAs for the new radiographic technique, the 6:1 grid ratio for the new technique represents an appropriate change in direction and magnitude.)

Problem # 5 (Finding a new kVp)

A radiographic technique calls for 400 mA at ½ second at 70 kVp, high-speed screens, 10:1 grid, a 40-inch FFD. What would the new kVp be to maintain equal radiographic densities if the new radiographic technique was 300 mA at 1 second, Par speed screens, 16:1 grid, at 50-inch FFD?

Solution

First, order both sets of techniques.

Original Technique	*New Technique*
400 mA	300 mA
½ sec	1 sec
70 kVp	? kVp
High Speed	Par Speed
40-inch FFD	50-inch FFD

Second, determine original mAs = 200 mAs.
Third, determine new mAs = 300 mAs (To be used in final solution in step six).
Fourth, convert 200 mAs at high speed to par speed = 400 mAs.
Fifth, convert 400 mAs at 40-inch FFD to 50-inch FFD = 625 mAs.
Sixth, convert 70 kVp at 625 mAs to 300 mAs = 80.5 kVp.

Problem # 6 (Finding a new kVp)

A radiographic technique calls for 50 mA at 2 seconds, 80 kVp, high-speed screens, 12:1 grid, and a 40-inch FFD. What would the new kVp be to maintain equal radiographic densities if the following radiographic technique were used instead: 300 mA at ½ second, slow-speed screens, 5:1 grid, and 50-inch FFD?

Solution # 1

First, order both sets of radiographic techniques.

Original Technique	*New Technique*
50 mA	25 mA
2 sec	5 sec
80 kVp	? kVp
High Speed	Slow Speed
12:1 grid	5:1 grid
40-inch FFD	50-inch FFD

Second, determine original mAs = 100 mAs.
Third, determine new mAs = 125 mAs (Hold until step #7).
Fourth, convert 100 mAs at high speed to slow speed = 400 mAs.
Fifth, convert 400 mAs at 12:1 to 5:1 = 160 mAs.
Sixth, convert 160 mAs at 40-nch FFD to 50-inch FFD = 250 mAs.
Seventh, convert 80 kVp at 250 mAs to 125 mAs = 92 kVp.

Solution # 2

Fifth, convert 80 kVp at 12:1 grid to 5:1 grid = 65 kVp.
Sixth, convert 400 mAs at 40-inch FFD to 50-inch FFD = 625 mAs.
Seventh, convert 65 kVp at 625 mAs to 125 mAs =

625 Mas at 65 kVp X .15 = 9.75 kVp + 65 kVp = 74.75 kVp at 312 mAs.
312 mAs at 74.75 kVp X .15 = 11.21 kVp + 74.75 kVp = 85.96 kVp at 156 mAs.
156 mAs at 85.96 kVp X .15 = 12.89 kVp + 85.96 kVp = 98.85 kVp at 78 mAs.

Interpolation required
78 mAs = 98.85 kVp
125 mAs = ? kVp
156 mAs = 85.96 kVp

Range of mAs = 156 mAs - 78 mAs = 78 mAs
Range of mAs up to 125 mAs = 125 - 78 = 47 mAs
Range of kVp = 98.85 kVp - 85.96 kVp = 12.89 kVp
Answer = $\frac{47}{78}$ of 12.89 kVp = 7.7 kVp

98.85 kVp - 7.7 kVp = 91.15 kVp (very close to 92 kVp)

Notice that in the above problem there were two methods for arriving at approximately similar answers of 91 or 92 kVp. However, the first solution provided a much easier and faster approach. Instead of converting 80 kVp from the 12:1 grid to the 5:1 grid, as was done in the second solution, the conversion of 400 mAs continued to the final solution involving conversion of 80 kVp at 250 mAs to 125 mAs = 92 kVp. These two approaches were illustrated so that the reader may realize that one solution principle may be easier than another even though they both arrive at similar answers. This may be of tremendous value when solving such problems where time is a critical factor.

Problem # 7 (Finding a new intensification screen)

A radiographic technique calls for the following: 100 mA at 1 second, 65 kVp, CBH (direct exposure), 40-inch FFD, and nongrid. What type of intensifying screen would be required in order to maintain similar radiographic densities if the technique was changed to 100 mA at 1/100 second, 75 kVp, 40-inch FFD, and 12:1 grid?

Solution

First, order both sets of radiographic techniques.

Original Technique	*New Technique*
100 mA	100 mA
1 sec	1/100 sec
65 kVp	75 kVp
40-inch FFD	40-inch FFD
nongrid	12:1 grid
CBH	? screen

Second, determine original mAs = 100 mAs.
Third, determine new mAs = 1 mAs (Hold until step #6)
Fourth, convert 100 mAs at 65 to 75 kVp = 50 mAs.
Fifth, 40-inch FFD is same, ignore.
Sixth, convert 50 mAs at CBH to 1 mAs at ? screen:

$$\frac{\text{Original Intensification Factor}}{\text{New Intensification Factor}} = \frac{\text{Original mAs}}{\text{New mAs}}$$

$$\frac{20}{\text{New IF}} = \frac{50 \text{ mAs}}{1 \text{ mAs}}$$

Cross Multiply

New IF (50 mAs) = 20 (1 mAs)

$$\text{New IF} = \frac{20\ (1\ \text{mAs})}{(50\ \text{mAs})}$$

New IF = .40

Now that the new Intensification Factor of .40 has been identified it must be matched as closely as possible to one of the known Intensification Factors. (See Table 17-1.) The closest intensification factor that would match is the .35 intensification factor for super-fast speed screens. Therefore, the correct screen type to be used for the new radiographic technique is super-fast speed screens.

Problem # 8 (Finding a new FFD)

A portable radiographic technique called for the following: 300 mA at 1/30 sec, 60 kVp, 10:1 grid, high-speed screens, at 40-inch FFD. What would the new FFD have to be in order to maintain similar densities if the following technique were used: 200 mA at 1 second, 80 kVp, nongrid, and CBH?

Solution

First, order both sets of radiographic techniques.

Original Techniques	*New Technique*
300 mA	200 mA
1/30 sec	1 sec
60 kVp	80 kVp
10:1 grid	nongrid
High Speed	CBH
40-inch FFD	? FFD

Second, determine original mAs = 10 mAs.
Third, determine new mAs = 200 mAs (Hold until step #7).
Fourth, convert 10 mAs at 60 to 80 kVp = 2.5 mAs.
Fifth, convert 2.5 mAs at 10:1 grid to nongrid = .5 mAs.
Sixth, convert .5 mAs at high speed to CBH = 20 mAs.
Seventh, convert 20 mAs at 40-inch FFD to 200 mAs at ? FFD = 126 inches.

$$\frac{\text{Original mAs}}{\text{New mAs}} = \frac{\text{Original FFD}^2}{\text{New FFD}^2}$$

$$\frac{20\ \text{mAs}}{200\ \text{mAs}} = \frac{40\text{-inch FFD}^2\ (1600\ \text{Inches})}{?\ \text{New FFD}^2}$$

Cross Multiply

$\text{New FFD}^2\ (20\ \text{mAs}) = 1600\ (200\ \text{mAs})$

$$\text{New FFD}^2 = \frac{1600\ (200\ \text{mAs})}{(20\ \text{mAs})}$$

$\text{New FFD}^2 = 16000$

$\text{New FFD} = \sqrt{16000}$

New FFD = 126 Inches

Radiographic Techniques Affecting Radiographic Density

To the novice, radiographic density problems may, at first, appear to be difficult. Identifying a strategy, however, and considerable practice at solving such problems will assist in building confidence. With these problems, there is little room for guessing the correct answer. Move on and learn that this type of exposure and imaging problem has a definite solution strategy and may be constructed with a variety of potential answer formats.

Problem # 1 (Finding the technique with the greatest density)

Which of the following radiographic techniques would result in the greatest amount of radiographic density?

	mA	*sec.*	*kVp*	*FFD*	*Screen*	*Grid*
A.	100	1.00	90	40″	High	5:1
B.	100	.50	60	40″	Slow	6:1
C.	300	1.50	70	50″	Super	8:1
D.	400	.25	80	50″	CBH	nongrid

Solution

First, determine all mAs values.
A. 100 mAs
B. 50 mAs
C. 450 mAs
D. 100 mAs

Second, select one kVp value and convert the other kVp values to it. Then, convert all mAs values accordingly.
A. 100 mAs at 90 kVp to 70 kVp = 400 mAs
B. 50 mAs at 60 kVp to 70 kVp = 25 mAs
C. 450 mAs at 70 kVp to 70 kVp = 450 mAs (no change)
D. 100 mAs at 80 kVp to 70 kVp = 200 mAs

Third, select one FFD value and convert the other FFD values to it with corresponding mAs changes.
A. 400 mAs at 40″ FFD (no change)
B. 25 mAs at 40″ FFD (no change)
C. 450 mAs at 50″ FFD to ? mAs at 40″ FFD = 288 mAs
D. 200 mAs at 50″ FFD to ? mAs at 40″ FFD = 128 mAs

Fourth, select one screen speed value and convert the other screen values to it with corresponding mAs changes.
A. 400 mAs at high to high = 400.00 mAs (no change)
B. 25 mAs at slow to high = 6.25 mAs
C. 288 mAs at super to high = 411.00 mAs
D. 128 mAs at CBH to high = 3.20 mAs

Fifth, select one grid ratio value and convert the other grid ratio values to it with corresponding mAs changes. (Converting all grid ratios to nongrid technique is another approach).
A. 400 mAs at 5:1 to 5:1 = 400.00 mAs (no change)
B. 25 mAs at 6:1 to 5:1 = 16.66 mAs
C. 288 mAs at 8:1 to 5:1 = 144.00 mAs
D. 128 mAs at nongrid to 5:1 = 256 mAs

Sixth, because the highest mAs value will represent the greatest amount of radiographic density, the obvious answer is 400 mAs as indicated by "A" above. This problem could have been the reverse situation in which the radiographer was required to select the radiographic technique that would result in the least amount of radiographic density. In this case the answer would have to be 16.66 mAs as indicated by "B" above. An entirely different approach might be to have the radiographer select from among the following answers with respect to the original problem from above. Having completed conversions up to the previous fifth

and final step, what would be the correct answer?
A. D will result in greater density than C (correct)
B. C will result in greater density than A
C. B will result in greater density than D
D. D will result in greater density than A

Problem # 2 (Finding the technique with the least density)

Which of the following radiographic techniques would result in the least amount of radiographic density?

	mA	*sec*	*Screen*	*FFD*	*kVp*	*Grid*
A.	100	.75	Slow	40″	60	nongrid
B.	200	.50	Par	40″	70	12:1
C.	1200	.01	High	72″	70	12:1
D.	50	.15	Super	55″	90	16:1

First, determine all mAs values =
A. 75 mAs
B. 100 mAs
C. 12 mAs
D. 7.5 mAs

Second, select one screen speed and convert all other screen speed values to this with respect to their mAs values.
A. 75 mAs at slow to slow = 75 mAs (no change)
B. 100 mAs at par to slow = 200 mAs
C. 12 mAs at high to slow = 48 mAs
D. 7.5 mAs at super to slow = 42.8 mAs (43 mAs)

Third, select one FFD value and convert all other FFD values to this with respect to their mAs values.
A. 75 mAs at 40″ FFD to 40″ FFD = 75 mAs (no change)
B. 200 mAs at 40″ FFD to 40″ FFD = 200 mAs (no change)
C. 48 mAs at 72″ FFD to 40″ = 14.8 mAs (15 mAs)
D. 43 mAs at 55″ FFD to 40″ FFD = 22.7 mAs (23 mAs)

Fourth, select one grid ratio and convert all other grid ratios to this value with respect to mAs.
A. 75 mAs at nongrid to 12:1 = 375 mAs
B. 200 mAs at 12:1 to 12: 1 = 200 mAs (no change)
C. 15 mAs at 12:1 go 12:1 = 15 mAs (no change)
D. 23 mAs at 16:1 to 12:1 = 19 mAs

Fifth, select one kVp value and convert all other mAs values.
A. 375 mAs at 60 kVp to 60 kVp = 375 mAs (no change)
B. 200 mAs at 70 kVp to 60 kVp = 400 mAs
C. 15 mAs at 70 kVp to 60 kVp = 30 mAs
D. 19 mAs at 90 kVp to 60 kVp = 152 mAs

Sixth, from the above step one may determine the least amount of radiographic density produced by locating the lowest mAs value which is 30 mAs as identified in item "C."

STUDY QUESTIONS

1. Given an exposure time of 1/100 second at 200 mA, calculate the mAs.

2. Given 1000 mA at 20 mAs, calculate exposure time.

3. Given 80 mAs at 1/40 second, calculate mA.

4. What are the relative intensification factors for calciu[m] tungstate intensification screens?

5. How are conversion factors for intensification screen[s] determined?

6. If 40 mAs is used for a par/medium screen, what would th[e] new mAs value be using high-speed screens in order to main[-]tain equivalent radiographic densities?

7. If the original radiation intensity was measured to be 100 roen[t]gens at 20 inches, what would the new intensity level b[e] at 30 inches?

8. If the original mAs was 100 at 20 inches, what would th[e] new mAs be at 30 inches in order to maintain equivalen[t] radiographic densities?

9. If the original mAs is 100 at 40 inches, what would the ne[w] distance need to be in order to double the mAs value an[d] still maintain equivalent radiographic densities?

10. Which of the following radiographic techniques produce th[e] least amount of heat units? (A) 30 mAs at 80 kVp, or (B) 10 mAs at 120 kVp.

11. A radiographic technique calls for a doubling effect o[n] radiographic density for the second exposure. The origina[l] technique is 200 mAs at 80 kVp. Which of the followin[g] techniques would be more beneficial to use for a doublin[g] density effect and state your reason why?
(A) 400 mAs at 80 kVp or (B) 200 mAs at 90 kVp

12. A radiographic technique calls for a halving effect o[n] radiographic density for the second exposure. The origina[l] technique is 200 mAs at 80 kVp. Which of the followin[g] techniques would be better to use for a halving density ef[-]fect and state your reason why?
(A) 100 mAs at 80 kVp or (B) 200 mAs at 70 kVp

13. List the grid conversion factors for mAs and kVp?

14. A radiographic technique with no grid calls for the use o[f] 100 mAs at 80 kVp. What would the new mAs value be i[f] a 16:1 grid were used?

15. A radiographic technique employs the use of a 16:1 grid a[t] 20 mAs and 80 kVp. What would be the new mAs value i[f] an 8:1 grid ratio were used?

16. A radiographic technique calls for the use of a 12:1 grid a[t] 3 mAs and 110 kVp. What would the new kVp be if a 6:1 grid were used instead?

17. Solve for the new exposure factor that will result in maintaining equivalent radiographic densities.

Original Technique	*New Technique*
400 mA	1200 mA
.01 sec	____ sec
80 kVp	90 kVp
12:1 grid	16:1 grid
high speed	super fast
40-inch FFD	56-inch FFD

18. Which of the following radiographic techniques would result in the least amount of radiographic density?

	mA	*sec*	*kVp*	*Screen*	*Grid*
A.	100	.10	90	High	12:1
B.	200	.10	80	Par	5:1
C.	300	.30	70	Slow	16:1
D.	400	.25	60	CBH	Nongrid
E.	600	.001	60	High	Nongrid

19. Which of the following radiographic techniques results in the longest scale of contrast even though they all contain essentially the same radiographic density?
A. 500 mAs at 50 kV
B. 250 mAs at 60 kV
C. 125 mAs at 70 kV
D. 62 mAs at 80 kV
E. 30 mAs at 90 kV

20. Which of the following radiographic techniques results in the shortest scale of contrast even though they all contain essentially the same radiographic density?
A. 100 mAs at 70 kV
B. 50 mAs at 80 kV
C. 25 mAs at 90 kV
D. 12 mAs at 100 kV
E. 6 mAs at 115 kV

BIBLIOGRAPHY

1. Bushong, Stewart C., *Radiologic Science for Technologists: Physics, Biology, and Protection*, C. V. Mosby Company, St. Louis, Missouri, 1980.
2. *Characteristics and Applications of X-Ray Grids*, Liebel-Flarsheim, Cincinnati, Ohio, 1983.
3. Hananel, Jeffrey I., *Solving Radiographic Technique Problems*, AMERAC, Los Angeles, 1979.
4. Lamel, David A., et. al.; *The Correlated Lecture Laboratory Series in Diagnostic Radiological Physics*, HHS Publication FDA 81-8150, U.S. Department of Health and Human Services, Bureau of Radiological Health, Rockville, Maryland, February, 1981.
5. Stevens, Matthew and Robert I. Phillips, *Comprehensive Review for the Radiologic Technologist*, 3rd ed.: C. V. Mosby Company, St. Louis, Missouri, 1977.
6. X-Ray Techniquiz, General Electric Company, Milwaukee, Wisconsin, 1965.

Appendix

Significant Events Leading Up to and Beyond the Discovery of X-Rays

600 B.C. Thales of Greece rubs amber and jet together and attracts straw.

1600 A.D. The physician, Gilbert, of Elizabeth's Court, described similarities and differences between the attraction of magnetic ore to iron and amber to light objects.

1672 Guerick makes the first electrical machine—a sulphur ball that could be whirled on bearings or rubbed with the hand.

1709 Hauksbee uses a glass plate to produce sparks.

1727 Schultz discovers that silver carbonate or chloride paste mixed with chalk becomes dark when exposed to light in a glass tube.

1733 du Fay shows that there are two distinct kinds of electricity, vitreous (repelling) and resinous (attracting).

1745 von Kleist and Masschenbroek at Leyden independently develop the primitive condenser which came to be known as the Leyden jar.

1750 Benjamin Franklin conducts famous kite experiments and names the vitreous charge—positive and the resinous charge—negative.

1766 Priestly, at Franklin's suggestion, demonstrates the absence of charge on the inside of a charged hollow sphere.

1775 Volta develops the electrophorus.

1780 Coulomb, Green's theory of electrostatics, and Galvani's discovery of battery action.

1787 Bennett invented the gold leaf electroscope.

1800 Nicholoson and Carlyle use new copper and zinc piles to cause electricity to flow through water.

1802 Wedgewood and Davy use paper coated with silver chloride to record silhouettes on glass by contact printing.

1816 Niepce makes a camera from a jewel box with a microscope lens to record nonpermanent images.

1819 Herschel discovers the action of sodium thiosulfate (hypo) on silver chloride.

1820 Oersted discovers that a wire carrying new battery currents affects a compass needle magnetically.

1826 Georg Ohm studies relationships between voltage, current, and resistance and establishes Ohm's law.

1831 Michael Faraday induces current in one coil by means of a current flowing through another.

1833 Faraday's laws of electrolysis are established.

1837 Reade uses sodium thiosulfate to dissolve remaining unexposed silver salts in photographic emulsion.

1838 Faraday passes electricity through gases and coins the terms "anode" and "cathode."

1839 Herschel coins the term, "photography."

Daguerre discovers the phenomenon of film development by accident.

1840 Talbot developed the "negative-positive" method of photography.

1845 Wilhelm Konrad Roentgen is born on March 27.

1847 Niepce de St. Victor coats glass with an albumen emulsion containing silver iodide and uses gallic acid as a development agent.

1850 Mercury exhaust pumps are invented.

1851 Heinrich Daniel Ruhmkorff develops the first efficient induction coil.

1870 Roentgen follows his mentor, Kundt, to Wurzburg as an assistant in physics.

Roentgen follows Kundt to Strassburg.

1871 Maddox invents the gelatin silver bromide dry plate.

1873 Burgess manufactures the first practical dry plate with a washed emulsion.

Vogel discovers that Burgess's plates, which were normally sensitive to blue and violet plates could be made sensitive to all colors by the addition of certain dyes.

1875 Roentgen, at age 30, is appointed Professor of Mathematics and Physics at the Agricultural College of Hohenheim.

1876 Roentgen returns to Strassburg at Kundts' request to assume the position of Associate Professor of Physics.

1879 George Eastman invented a plate-coating machine for the first manufacturing of coated glass plates.

Roentgen is called to Giessen as Professor and Director of the Institute of Physics.

1888 Roentgen is called to the University of Wurzburg, Bavaria, to assume the position of Director of the Physical Institute.

Crookes focuses cathode rays using a concave-shaped cathode and demonstrates cathode rays produce fluorescence in various salts and that narrow beams impinged on fluorescent screens can be deflected by a magnet.

1894 Leonard passes cathode rays into the air through a window of aluminum foil and ponders the possibility that there may be some form of invisible light.

1895 Roentgen discovers on November 8, that when cathode rays strike the wall of an exhausted tube they produce heat, fluorescence, and a new form of invisible light (X-Rays) capable of penetrating many substances opaque to ordinary light.

Roentgen presents for publication in the Annals of the Wurzburg Physical Medical Society on December 28, preliminary paper on x-rays entitled, *Uber Eine Neve Art von Strahlen*.

Arthur W. Fuchs is born on 1 April.

1896 Roentgen presents his preliminary paper before the Wurzburg Physical Medical Society on January 23.

The first medical x-ray examination in the United States is conducted on a patient in the physics laboratory at Dartmouth College on February 3.

Roentgen submits a second paper on X-Rays to the Wurzburg Physical Medical Society on March 9.

Cannon at Harvard fed bismuth to a goose and later to cats as a contrast medium.

Lieutenant Colonel Giuseppe Alvaro of Italy is the first to use x-rays in a military capacity for foreign body localization in the Ethiopian Campaign.

J. Daniel is first to report x-ray epilation.

Wolfram Conrad Fuchs establishes the first commercial x-ray laboratory in Chicago.

Thomas Edison develops a calcium tungstate intensifying screen which was first used by Professor Pupin of Columbia University.

1897 Roentgen submits a third communication on X-Rays to the Royal Prussian Academy of Science, in Berlin, entitled, *On Further Observation of the Characteristics of the X-Rays*.

Wagner Company makes the glass plate static machine.

1898 First use of the x-rays by the United States Army in the Spanish American War.

Thomas Edison develops the fluoroscope usin barium platinocyanide.

1899 Cannon at Harvard uses bismuth on humans as a co trast medium.

1900 Roentgen is made a member of the Munich Academ of Science.

1901 Roentgen is awarded the Nobel Prize in Physics.

American Roentgen Ray Society is organized.

1904 C. M. Dally is the first radiation fatality.

Charles L. Leonard demonstrates that by exposin two glass x-ray plates with the emulsion surface together, exposure time was reduced and the imag was considerably enhanced.

1905 First pyelograms and cystograms were made usin colloidal silver.

1907 Homer Clyde Snook develops the interrupterles transformer for static and coil x-ray machines whic exceeded voltage requirements for Crookes tubes

M. K. Kassabian publishes his classic textbook, *Roentgen Rays and Electro-Therapeutics*.

Wolfram C. Fuchs, a roentgen pioneer, dies as result of effects suffered from severe x-ray dermatitis

1910 M. K. Kassabian, a roentgen pioneer, dies in Philadelphia on July 14 as a result of metastatic malignancy o the axilla due to severe x-ray dermatitis of the hand that was contracted early on in his career.

1911 Gustav Bucky develops the antiscatter diaphragm.

1912 Dr. W. D. Coolidge develops the hot cathode x-ra tube.

Vabor injects air as a contrast medium into the abdominal cavities of experimental animals.

1913 Kodak introduces "Eastman X-Ray Film," a cellulose nitrate base coated on one side.

1914 World War I halts the supply of Belgium glass use on the production of glass plates in America.

1915 Radiological Society of North America is organized

1916 Hollis Elmer Potter develops the reciprocating Bucky

American Radium Society is organized.

1917 Dr. W. D. Coolidge develops the radiator-type ho cathode ray tube.

The American Roentgen Ray Society recommend that military schools of roentgenology be establishe among civilian institutions at nine American cities fo the training of reserve corps and active duty medica officers (physicians) in preparation for World War I

Formal training of enlisted men as x-ray manipulator begins at the United States Army Medical School i

Washington D.C., Ft. Riley, Kansas, and Camp Greenleaf, Georgia.

First U.S. Army X-Ray Manual is hurriedly prepared for the training of military roentgenologists and x-ray manipulators.

1918 Sodium iodide replaces colloidal silver as a safer and cheaper contrast medium.

United States Army X-Ray Manual and its *Extract* are published in October. Lieutenant Colonel John Shearer, Ph.D., writes the chapter on x-ray physics.

Eastman Kodak develops ''Eastman Dupli-Tized X-Ray Film,'' a double coated x-ray film that could be used with intensifying screens.

1919 Dandy introduces air as a contrast medium into the ventricular system through the skull.

Roentgen's wife, Berta, dies on October 31.

H. F. Waite develops the oil-immersed transformer and x-ray tube.

1920 American Association of Radiological Technicians is organized with Ed. C. Jerman elected as first president.

1921 Secard and Forestien develop lipidol as a contrast medium for injection into the bronchial tree and central body cavities.

1923 Roentgen dies from rectal cancer on February 10.

American Registry of Radiological Technicians is organized by the American Roentgen Ray Society and Ed. C. Jerman is appointed as first certifying examiner.

Classification Act is passed by Congress which defines x-ray technicians as subprofessionals who work under the radiologist.

American College of Radiology is organized.

*Almost every phase of diagnostic x-ray service reaches the stage of development which is recognized by today's standards as modern.

1924 Kodak introduces ''Eastman Safety Dupli-Tized X-Ray Film,'' the first x-ray film with an acetate safety base.

1928 Ed. C. Jerman publishes, *Modern X-Ray Technic.*

LeRoy Sante publishes textbook, *Manual of Radiological Technique.*

1929 First publication of journal, *X-Ray Technician*, by American Society of X-Ray Technicians.

Introduction of the rotating anode.

1930 First publication of journal, *Radiography and Clinical Photography*, by Kodak with Arthur W. Fuchs as editor.

* This chronology focuses primarily on events leading up to and immediately after the discovery of x-rays. The modern period, 1923 to the present, has seen tremendous advances in automatic processing, x-ray film and intensifying screen construction, along with incredible advances in every area of radiologic technology including magnificent discoveries that have lead to the development of computed tomography, ultrasonography and magnetic resonance imaging, to name a few.

Index

accidental motion, 109
actinography, 4
actual focal spot, 64
air-gap techniques, 59
alignment (see tube-object-film alignment)
American Roentgen Ray Society, 5
American Radiologic Society of North America, 5
American Registry of Radiologic Technicians, 5-6
American Association of Radiologic Technicians, 5
anode, 49
aperature diaphragms, 75
Army, U.S., 5, 6, 10
artifacts, 35-36
attenuation, 69, 70
automatic exposure timer (AET), 112-113
automatic film processing, 116-117

beam restrictors
- collimators, 71, 76-78
- contrast, 77-78
- density, 77-78
- diaphragms, 75
- extension cylinders, 76
- flare cones, 75-76
- "framing", 83
- recorded detail, 79
- role in radiography, 75
- shape distortion, 79
- size distortion, 79
- types, 75-76

"bloom" effect, 55
body classifications, 87-88
body-section, 110
Bucky, Gustave, 93-94
Bucky-Potter diaphragm (see also grid), 96, 98
- alignment, 82
- history, 7, 8

calcium tungstate screens, 9
cathode, 49
characteristic curve, 10
Classification Act, 6
coil-type x-ray machines, 6
collimators, 71, 76-78
compensating filter, 71-72
contrast media, 63
contrast
- beam restrictors, 77-78
- conversion problems, 121-131
- contrast, 131
- defined, 17, 18, 21
- density, 129-130
- distance, 60
- film, 24
- film processing, 117-118
- filtration, 72
- focal spot size, 65
- grids, 98, 100, 125-129
- intensifying screens, 105-106, 122-123, 128-129
- kVp, 44-47
- kVp/mAs, 124-126
- latitude, 23
- mAs, 55
- mAs/FFD, 123-124, 129-131
- motion, 110-111
- patient, 91-92
- problems, 131
- subject, 18, 21-22

Coolidge, W.D., 7
Crookes (Hittorf) tube, 3, 8
cutoff, grid, 97-98

definition, recorded, 27-28
densitometry, 18-19
density
- beam restrictors, 77-78
- compression, 21
- defined, 18
- diagnostic viewing range, 20
- distance, 59-60

film processing, 117-118
filtration, 72
focal spot size, 65
grids, 98, 99
intensifying screens, 104-105
kVp, 44
mAs, 53-55
motion, 110
part thickness, 21
patient, 90-91
problems, 129-130

detail (see recorded detail)
developer solution, 11-12, 115, 117, 118
diagraphy, 4
diaphrams, 75
distance
air gap techniques, 59-60
contrast, 60
density, 59
inverse square law, 58
mAs-Distance formula, 58
problems, 123-124, 129
recorded detail, 60-61
role in radiography, 57
shape distortion, 61
size distortion, 61

distortion (see size and shape distortion)

Eastman Kodak, 7, 8
effective focal spot, 64
electro-skiagraphy, 4
electrography, 4
elongation, 32, 33, 84-85
evaluation, film, 39-40
exposure latitude, 47
extension cylinders, 76, 77

fading, safelight, 116
FFD (see distance)
Fifteen Percent Rule, 124-125
filament, 49-50, 64
filament "blooming", 65
film processing
automatic, 116-117
considerations, 115-117
contrast, 117-118
density, 117
developer, 115
fixer, 12, 115, 117
manual, 116
recorded detail, 118
safelights, 115-116
shape distortion, 118
size distortion, 118
temperature, 117-118

filter
compensating, 71-72
equalizing, 71
safelight, 115-116
wedge, 71-72

filtration
added, 70
collimator, 70, 71
contrast, 72
density, 72
half-value layer, 71
inherent, 70
recorded detail, 72
role in radiography, 69
shape distortion, 72
size distortion, 72

fixer 12, 115, 117
flare cones, 75-76
"floating" tabletops, 82-83
fluorography, 4
focal film distance (see distance)
focal spot size
contrast, 65
control, 64
density, 65
measuring tools, 64
recorded detail, 65
role in radiography, 63
shape distortion, 66
size distortion, 66
structure, 63

focus, grid, 95
focusing cup, 64, 65
fog, 44
foreshortening, 32, 33, 84-85
"framing", 83
Fuchs, Wolfram, 4
Fuchs, Arthur, 4

gaseous x-ray tube, 6
General Electric Company, 11
geometric unsharpness, 28
graininess, film, 37-38
grid
applications, 96
Bucky, 96-97
construction, 93-94
contrast, 98, 100
conversation problems, 125-129
criss-cross, 94-95
crossed, 94-95
cross-hatch, 94-95
cutoff, 97
density, 98, 99
focus, 95-96

history, 93-94
linear, 94-95
lines, 96, 97
parallel (unfocused), 94
radius, 95, 97
ratio, 95
recorded detail, 99
shape distortion, 99
size distortion, 99
stationary, 96
strip thickness, 95
terminology, 95
types, 95-96
wafer, 95
Gurney-Mott Theory, 18

H & D curves, 117-118
half-value layer, 71
"heel" effect, 63-64
Hennecart, M., 4
heterogeneous, 69
high tube, 6-7
history (also see Appendix)
commercialization, 4
film, 7-8
processing, 11-13
profession, 4-6
protection, 13-15
Roentgen, Wilhelm, 2-4
technique, 8-11
x-ray machine, 6
x-ray tubes, 6-7
hot cathode tube, 9
"hot-lighting", 20-21
Hurter & Driffield, 19

illuminators, 38
immobilization, 113
intensification factor, 29
intensifying screens
contrast, 105-106
conversion problems, 122-123, 128-129
density, 104-105
efficiency, 102
phosphor, 101
protective coating, 101-102
quantum mottle, 103-104
recorded detail, 106-107
reflective layer, 101
role in radiography, 101-104
shape distortion, 107
size distortion, 107
"speeds", 102
undercoat, 101
unsharpness, 28, 103

inverse square law, 58
involuntary motion, 109
ixography, 4

Jerman, Ed., 5-6, 8, 10-11

Kassabian, Dr. Mirhan, 6, 13-14
kathography, 4
kilovoltage (see kVp)
Kundt, A., 3
kVp
contrast, 44-47
density, 44
exposure latitude, 47
recorded detail, 47
role in radiography, 43
rule of 10, 44
shape distortion, 47
size distortion, 47
kVp/mAs conversion problems, 124-126

laminography, 110
latitude
defined, 23
exposure, 23-24, 47
film, 24, 25
lead diaphragms, 75, 76
Line Focus Principle, 64
low tube, 6-7

macroradiography, 66
magnification
factor, 31
radiography (also see size distortion), 66
mammography, 71
manipulators, 5
manual film processing, 116
mAs
contrast, 55
density, 53-55
-Distance formula, 58-59
problems, 123-124, 129-131
recorded detail, 55
role in radiography, 49-53
shape distortion, 55
size distortion, 55
Merrigan, M. Beatrice, 6
milliampere-seconds, 49-56
molybdenum, 7
motion, 109-114
motion
accidental, 109
contrast, 110-111
density, 110
involuntary, 109

physiological, 109
recorded detail, 109-110
role in radiography, 109-110
shape distortion, 113
size distortion, 113
tomography, 110, 111
unsharpness, 30
mottle, 35, 36-38
mottle, quantum, 29, 36-37, 103-104

nephrography, 110
new photography, 4
noise
artifacts, 35-36
defined, 35
film graininess, 37-38
mottle, 35, 36-38
radiographic, 35
nonscreen filmholders, 102-103

object film distance (OFD), 58-62
optical focal spot, 64

Paris Academie de Medecine, 5
patient
body types, 87
contrast, 91-92
density, 90-91
pathology, 89-90
recorded detail, 92
role of tissue quality, 87-90
shape distortion, 92
size distortion, 92
tissue opacity, 88
penumbra, 28, 103
penumbral effect, 9
phosphors, screen, 28-29, 101-102
physiological motion, 109
plainography, 110
platino-cyanide of barium, 9
polychromatic, 69
portable radiography, 83-84
Potter, Dr. Hollis, 96
problems, conversion, 121-130
processing (see film processing)
psyknoscopy, 4

quantum mottle, 29, 36-37, 103-104

radiation burns, 13
radiation protection, 13
radiodermatitis, 4, 13
radiographic technique, 10-11
radiography, 4
radiolucence, 22, 92
radiopacity, 22, 92
rare-earth screens, 28
rating chart, tube, 52, 54
ratio, grid, 95
reciprocity, 18
recorded detail
beam restrictors, 79
defined, 28
distance, 60-61
film processing, 118
filtration, 72
focal spot size, 65-66
grids, 99
intensifying screens, 28, 106-107
kVp, 47
mAs, 55
motion, 111-113
patient, 91
reflective layer, 101
repeating radiographs, criteria, 40
Roentgen, Wilhelm, 2-4, 7
Roentgen Society of America, 4
Roentgen-ray burns, 4
roentography, 4
Rule of 10, 44, 124

safelights, 115-116
screen speed, 28
screen mottle, 37
screens (see intensifying screens)
seconds, 49-54
sensitometric curve, 19
shadowgraphy, 4
shape distortion
beam restrictors, 79
defined, 28, 31, 84
distance, 61
elongation, 32, 33, 84-85
film processing, 118
filtration, 72
focal spot size, 66
foreshortening, 32, 33, 84-85
grids, 99
intensifying screens, 107
kVp, 47
mAs, 55
motion, 113
patient, 92
tube-object-film alignment, 84-85
Shearer, John, 10
silver, black metallic, 18
size distortion
beam restrictors, 79
defined, 28, 30
distance, 61

film processing, 118
filtration, 72
focal spot size, 66
grids, 99
intensifying screens, 107
kVp, 47
magnification factor, 31
mAs, 55
motion, 113
patient, 92
skiagraphy, 4
skiography, 4
skotography, 4
Snook transformer, 7, 8
Snook tube, 6
source image receptor distance (see FFD)
Spanish-American War, 5
spark gaps, 8
speed, intensifying screen, 102
static machines, 6
structure mottle, 37

target, 49
target film distance (see FFD)
technique charts, 9
thermionic emission, 49, 64
time, exposure, 49-54
tissue quality, 87-90
tomography, 110
true distortion, 31
tube-object-film alignment
Bucky, 82
floating tabletops, 82
geometric components, 81-82
horizontal central ray, 83
portable radiography, 83
shape distortion, 84-85
tungsten, 7

umbra, 28
unsharpness (also see recorded detail)
geometric, 28
screen, 28
motion, 30

vacuum, 7
viewboxes, 38
viewing conditions, 38

wafer grids, 95-96
wedge filter, 71
Wisconsin Test Cassette, 71
Wurzburg, U. of, 3

x-ray plates, 8
x-ray history (see history)